WORLD Bₗ

By

H. G. WELLS

... may knowledge never
stand in the way of truth!

— with the best wishes

Aneta, Alan and Sophie

Bury 23/09/2020

British Library Cataloguing-in-Publication Data
A catalogue record for this book is available from
the British Library

CONTENTS

H. G. WELLS

Herbert George Wells was born in Bromley, England in 1866. He apprenticed as a draper before becoming a pupil-teacher at Midhurst Grammar School in West Sussex. Some years later, Wells won a scholarship to the School of Science in London, where he developed a strong interest in biology and evolution, founding and editing the *Science Schools Journal*. However, he left before graduating to return to teaching, and began to focus increasingly on writing. His first major essay on science, 'The Rediscovery of the Unique', appeared in 1891. However, it was in 1895 that Wells seriously established himself as a writer, with the publication of the now iconic novel, *The Time Machine*.

Wells followed *The Time Machine* with the equally well-received *War of the Worlds* (1898), which proved highly popular in the USA, and was serialized in the magazine *Cosmopolitan*. Around the turn of the century, he also began to write extensively on politics, technology and the future, producing works *The Discovery of the Future* (1902) and *Mankind in the Making* (1903). An active socialist, in 1904 Wells joined the Fabian Society, and his 1905 book *A Modern Utopia* presented a vision of a socialist society founded on reason and compassion. Wells also penned a range of successful comic novels, such as *Kipps* (1905) and *The History of Mr Polly* (1910).

Wells' 1920 work, *The Outline of History,* was penned in response to the Russian Revolution, and declared that world would be improved by education, rather than revolution. It made Wells one of the most important political thinkers of the twenties and thirties, and he began to write for a number of journals and newspapers, even travelling to Russia to lecture Lenin and Trotsky on social reform. Appalled by the carnage of World War II, Wells began to work on a project dealing with the perils of nuclear war, but died before completing it. He is now regarded as one of the greatest science-fiction writers of all time, and an important political thinker.

PREFACE

THE papers and addresses I have collected in this little book are submitted as contributions, however informal, to what is essentially a scientific research. But it is a research in a field to which scientific standing is not generally accorded and where peculiar methods have to be employed. It is in the field of constructive sociology, the science of social organisation. This is a special sub-section of human ecology, which is a branch of general ecology, which again is a stem in the great and growing cluster of biological sciences. It stands with palaeontology at the opposite pole to experimental biology; hardly any verificatory experiment is possible and no controls. It is a science of pure observation, therefore, of analysis and of search for confirmatory instances. On the one hand it passes, without crossing any definite boundaries, into historical science proper, into the analysis of historical fact, that is, and on the other into the examination of such matters as geographical (and geological) conditions and the social consequences of industrial processes.

Human ecology surveys the species *Homo sapien* as a whole in space and time; sociology is that part of the survey which concerns itself with the interaction and interdependence of human groups and individuals. It is hardly to be distinguished from social psychology. There has been an enormous increase in the intensity and scope of human interaction and interdependence during the past half-million years or more. Communities and what one may call ranges of reaction, have enlarged and continue to enlarge more and more rapidly towards a planetary limit. The human intelligence is involved in this enlargement and it is too deeply concerned with its role in the process, to observe it with the detachment it can maintain towards the facts, for example, of astronomy or crystallography. Constructive sociology has to bring not only the study of conduct but an irresistible element of purpose into its problems. Human beings are not simply born or thrown together into association like a swarm of herrings. They keep together with a sense of collective activities and common ends, even if these ends are little more than mutual aid, protection and defence.

Throughout the whole range of ecology we study the adaptation of living species to changing environments, but outside the human experience these adaptations are generally made unconsciously by the natural selection of mutations and variations. These adaptations are inherited. They are either successful and the species is modified and survives, or it perishes. In the cerebral animals, however, natural selection is supplemented by very considerable individual adaptability. Memories and habits are established in each generation which fit individuals to the special circumstances of their own generation. They are adaptations which perish with the individual. Such creatures learn; they are educable creatures; dogs, cats, seals, elephants for example learn and the next generation has, if necessary, to learn the old lesson all over again or a different lesson. In the human being there is an unprecedented extension of educability; not only is learning developed to relatively immense proportions, but it is further supplemented by curiosity, precept and tradition. In such a slow-breeding creature as man educational adaptation is, beyond all comparison, a swifter process than genetic adaptation. His social life, his habits, have changed completely, have even undergone reversion and reversal, while his heredity seems to have changed very little if at all, since the late Stone Age. Possibly he is more educable now and with a more prolonged physical and mental adolescence.

The human individual is born now to live in a society for which his fundamental instincts are altogether inadequate. He has to be educated systematically for his social role. The social man is a manufactured product of which the natural man is the raw nucleus. In a world of fluctuating and generally expanding communities and ranges of reaction, the science of constructive sociology seeks to detect and give definition to the trends and requirements of man's social circumstances and to study the possibilities and methods of adapting the natural man to them. It is the science of current adaptations. It has therefore two reciprocal aspects; on the one hand it has to deal with social organisations, laws, customs and regulations, which may either be actually operative or merely projected and potential, and on the other hand it has to examine the education these real or proposed social organisations require. These two aspects are inseparable, they need to fit like hand and glove. Plans and theories of social structure and plans and theories of education are the outer and inner aspects of the same thing. Each necessitates the other. Every social order must have its own distinctive process of education.

In the past this imperative association of education and social structure was not recognised so clearly as it is at the present time. Communities would grow up and not change their mental clothes until they burst out of them. Ideas would change and disorganise institutions. For the past twenty-six centuries, and particularly, and much more definitely, during the last three, there has been a very great expenditure of mental energy upon the statement in various terms and metaphors, as theologies, as religions, socialisms, communisms, devotions, loyalties, codes of behaviour and so on, of the desirable and necessary form of human adaptation to new conditions of association. From the point of view of constructive sociology, or to coin a hideous phrase, human adaptology, all these efforts, although not deliberately made as experiments, are so much experience and working material, and though almost all of them have involved special teachings (doctrines) the need for a close interlocking of training and teaching with the social order sought, though always fairly obvious, has never been so fully realised as it is today. The new doctrines were often only subconsciously linked to the new needs. The idea, for instance, of a universal God replacing local gods ensued upon the growth of great empires, but it was not explicitly related to the growth of great empires. The connection was not plainly apparent to men's minds. In the looser, easier past of our species, there has never been such a close interweaving of current usage and practices with instruction and precept as we are now beginning to feel desirable. The reference of one to the other was not direct. Now education becomes more and more definitely political and economic. It must penetrate deeper and deeper into life as life ceases to be customary and grows more and more deliberately planned and adjusted. The need for lively and continuous invention in constructive sociology and for an animated and animating progressive education correlated with these innovations, has hardly more than dawned on the world. The urgency of adaptation has still to be grasped.

Throughout the nineteenth century certain systems

of adaptive ideas spread throughout the world to meet the requirements of what was recognised with increasing understanding as a new age. Mechanism was altering both the fundamental need for toil and the essential nature of war. The practical and cynically accepted need for labouring classes and subject-peoples, was dissolving quietly out of human thought— though it still exists in the minds of those who employ personal servants. Means of intercommunication and mutual help and injury have developed amazingly. A mechanical unification of the world has been demanding (and still demands) profound moral and ideological readjustments. It is, for example, being realised, slowly but steadily, that the fragmentary control of production and trade through irresponsible individual ownership gives quite lamentably inadequate results, that the whole property-money system needs revision very urgently, and that the belated recrudescence of sentimental nationalism largely through misguided school-teaching and newspaper propaganda, is becoming an increasing menace to world welfare. The old ideological equipments throughout the world are misfits everywhere. Mental and moral adaptation is lagging dreadfully behind the change in our conditions. A great and menacing gulf opens which only an immense expansion of teaching and instruction can fill.

In the field of sociology it is impossible to disentangle social analysis from literature, and the criticism of the social order by Ruskin, William Morris and so forth was at least as much a contribution to social science as Herbert Spencer's quasi-scientific defence of individualism and the abstractions and dogmas of the political economists. The biological sciences did not spread very easily into this undeveloped region. It was a hinterland of novel problems and possibilities. Even today proper methods of study in this field have still to be fully worked out and brought into association. It has had to be explored by moral and religious appeals, by utopias and speculative writings of a quality and texture. very unsatisfying to scientific workers in more definite fields. It is still subject to irruptions of a type that the normal scientist of today finds highly questionable. Poets even and seers have their role in this experimentation. But economics and sociology can only be made hard sciences by eliminating much of their living content. Knowledge has to be attained by any available means. Inquirers cannot be limited to passable imitations of the methods followed in other fields. It may be doubted if constructive sociology and educational science can ever be freed from a certain literary, aesthetic and ethical flavouring. We have to assume certain desiderata before we can get down to effective, applicable work. Yet it does seem possible to state the problem of adaptation in practical scientific terms.

It was not realised at first, and it is still not fully realised, how vague and unsuitable for immediate application the generous propositions of Socialism and World Peace remain, until further intensive and continuous research and elaboration have been undertaken. It is widely assumed that to profess Socialism or Pacifism implies the immediate undertaking of vehement political activities unencumbered by further thought. But the profession of Socialism or World Peace should commit a man to nothing of the sort. Socialism and World Peace are hardly more than sketches of the general frame of adaptation of which our species stands in need. "We are all socialists nowadays," but all the same there is very little really efficient working socialism.

"All men are brothers "—we have echoed that since the days of Buddha and Christ—but Spain and China are poor evidence of that fraternity. We know we want these things quite clearly, but we have still to learn how they are to be got.

Man reflects before he acts, but not very much; he is still by nature intellectually impatient. No sooner does he apprehend, in whole or in part, the need of a new world, than, without further plans or estimates, he gem into a state of passionate aggressiveness and suspicion and sets about trying to change the present order. There and then, he sets about it, with anything that comes handy, violently, disastrously, making the discordances worse instead of better, and quarrelling bitterly with any one who is not in complete accordance with his particular spasmodic conception of the change needful. He is unable to realise that when the time comes I to act, that also is the time to think fast and hard. He will not think enough.

There has been, therefore, an enormous waste of human mental, moral and physical resources in premature revolutionary thrusts, ill-planned, dogmatic, essentially unscientific reconstructions and restorations of the social order, during the past hundred years. This was the inevitable first result of the discrediting of those old and superseded mental adaptations which were embodied in the institutions and education of the past. They discredited themselves and left the world full of problems. The idea of expropriating the owners of land and industrial plant, for instance (Socialism), long preceded any deliberate attempt to create a Competent Receiver. Hysterical objection to further research, to any sustained criticism, has been, and is still characteristic of nearly all the pseudo-constructive movements of our time, culminating in projects for a "seizure of power" by some presumptuous oaf or other. The meanest thing in human natures is the fear of responsibility and the craving for leadership. "Right" dictators there are and "Left" dictators, and in effect there is hardly a pin to choose between them. The important thing about them from our present point of view, is that fear-saturated impatience for guidance, which renders dictatorships possible. First there comes a terrifying realisation of the limitless uncontrolled changes now in progress, then wild stampedes, suspicions, mass murders and finally *mus ridiculus* the Hero emerges, a poor single, silly, little human cranium held high and adorned usually with something preposterous in the way of hats. "He knows," they cry. "Hail the Leader!" He acts his part; he may even believe in it. And for quite a long time the crowd will refuse to realise that not only is nothing better than it was before, but that change is still marching on and marching at it—as inexorably as though there were no Leaders on the scene at all. Between the extremes of Right and Left hysteria, there remains a great under-developed region in the world of political thought and will, that we may characterise as "do-nothing democracy". Out of the sudden realisation of its do-nothingness arise these psychological storms which give gangster dictators their opportunities. It is only gradually that people have come to realise that current democratic institutions are a very poor, slow and slack method of conducting human affairs which need an exhaustive revision, and that when one has declared oneself Anti-Fascist, Anti-Communist or both, one has still said precisely nothing about the government of the world. One is brought back to the unsolved problem of the Competent Receiver. It exercised Plato. It has been intermittently revived and neglected ever since.

It is an intricate and difficult problem. To that I can testify because for more than half my life it has been my main preoccupation. The attack on this problem is, to begin with, a task to be done in the study and in the unhurried and irresponsible spirit of pure inquiry. As the attack gathers confidence a taint of propaganda may easily infect it, but the less that constructive sociology is propagandist, the higher will be its scientific standing and the greater its ultimate usefulness to mankind. The application of the results of its researches is another business altogether, the business of the statesman, organiser and practical administrator. And in spite of the paucity of disinterested explorers in this region of speculation and analysis, and in spite of the lack of effective discussion and interchange in this field (due mainly, I think, to the inadequate recognition of its immense scientific importance which forces its workers so often into a hampering association with politically active bodies) there does seem to be a growing and spreading clarification of the realities of the human situation. It is becoming apparent that the real clue to that reconciliation of freedom and sustained initiative with the more elaborate social organisation which is being demanded from us, lies in raising and unifying, and so implementing and making more effective, the general intelligence services of the world. That at least is the argument in this book. The missing factor in human airs, it is suggested here, is a gigantic and many-sided educational renascence. The highly educated section, the finer minds of the human race are so dispersed, so ineffectively related to the common man, that they are powerless in the face of political and social adventurers of the coarsest sort. We want a reconditioned and more powerful Public Opinion. In a universal organisation and clarification of knowledge and ideas, in a closer synthesis of university and educational activities, in the evocation, that is, of what I have here called a World Brain, operating by an enhanced educational system through the whole body of mankind, a World Brain which will replace our multitude of unco-ordinated ganglia, our powerless miscellany of universities, research institutions, literatures with a purpose, national educational systems and the like; in that and in that alone, it is maintained, is there any clear hope of a really Competent Receiver for world affairs, any hope of an adequate directive control of the present destructive drift of world affairs. We do not want dictators, we do not want oligarchic parties or class rule, we want a widespread world intelligence conscious of itself To work out a way to that World Brain organisation is therefore our primary need in this age of imperative construction.

It is an immense undertaking but not an impossible undertaking. I do not think there is any insurmountable obstacle in the way to the production of such a ruling World Brain. There are favourable conditions for it, encouraging precedents and a plainly evident need. The various lectures, addresses and papers, collected here, few, thin and sketchy though they may seem, are all in their scope and measure contributions to this urgent research.

H.G. Wells

I. WORLD ENCYCLOPAEDIA

LECTURE DELIVERED AT THE WEEKLY EVENING
MEETING OF THE ROYAL INSTITUTION OF GREAT
BRITAIN, NOVEMBER 20TH, 1936

MOST of the lectures that are given in this place to this audience are delivered by men of very special knowledge. They come here to tell you something you did not know before. But tonight I doubt if I shall tell you anything that is not already quite familiar to you. I am here not to impart facts but to make certain suggestions. And there is no other audience in the world to which I would make these suggestions more willingly and more hopefully than I do to you.

My particular line of country has always been generalisation and synthesis. I dislike isolated events and disconnected details. I really hate statements, views, prejudices and beliefs that jump at you suddenly out of mid-air. I like my world as coherent and consistent as possible. So far at any rate my temperament is that of a scientific man. And that is why I have spent a few score thousand hours of my particular allotment of vitality in making outlines of history, short histories of the world, general accounts of the science of life, attempts to bring economic, financial and social life into one conspectus and even, still more desperate, struggles to estimate the possible consequences of this or that set of operating causes upon the future of mankind. All these attempts had profound and conspicuous faults and weaknesses; even my friends are apt to mention them with an apologetic smile; presumptuous and preposterous they were, I admit, but I look back upon them, completely unabashed. Somebody had to break the ice. Somebody had to try out such summaries on the general mind. My reply to the superior critic has always been—forgive me—"Damn you, do it better".

The least satisfactory thing about these experiments of mine, so far as I am concerned, is that they did not at once provoke the learned and competent to produce superior substitutes. And in view of the number of able and distinguished people we have in the world professing and teaching economic, sociological, financial science, and the admittedly unsatisfactory nature of the world's financial, economic and political affairs, it is to me an immensely disconcerting fact that the Work, Wealth and Happiness of Mankind which was first published in 1932 remains—practically uncriticised, unstudied and largely unread—the only attempt to bring human ecology into one correlated survey.

Well, I mention this experimental work now in order that you should not think I am throwing casually formed ideas before you tonight. I am bringing you my best. The thoughts I am setting out here have troubled my mind for years, and my ideas have been slowly gathering definition throughout these experiments and experiences. They have interwoven more and more intimately with other solicitudes of a more general nature in which I feel fairly certain of meeting your understanding and sympathy.

I doubt if there is anybody here tonight who has not given a certain amount of anxious thought to the conspicuous ineffectiveness of modern knowledge and—how do I call it?—

trained and studied thought in contemporary affairs. And I think that it is mainly in the troubled years since 1914 that the world of cultivated, learned and scientific people of which you are so representative, has become conscious of this ineffectiveness. Before that time, or to be more precise before 1909 or 1910, the world, our world as we older ones recall it, was living in a state of confidence, of established values, of assured security, which is already becoming now almost incredible. We had no suspicion then how much that apparent security had been undermined by science, invention and sceptical inquiry. Most of us carried on into the War, and even right through the War, under the inertia of the accepted beliefs to which we had been born. We felt that the sort of history that we were used to was still going on, and we hardly realised at all that the war was a new sort of thing, not like the old wars, that the old traditions of strategy were disastrously out of date, and that the old pattern of settling up after a war could only lead to such a thickening tangle of evil consequences as we contemplate today. We know better now.

Wiser after the events as we all are, few of us now fail to appreciate the stupendous ignorance, the almost total lack of grasp of social and economic realities, the short views, the shallowness of mind, that characterised the treaty-making of 1919 and 1920. I suppose Mr. Maynard Keynes was one of the first to open our eyes to this worldwide intellectual insufficiency. What his book, The Economic Consequences of the Peace, practically said to the world was this: These people, these politicians, these statesmen, these directive people who are in authority over us, know scarcely anything about the business they have in hand. Nobody knows very much, but the important thing to realise is that they do not even know what is to he known. They arrange so and so, and so and so must ensue and they cannot or will not see that so and so must ensue. They are so unaccustomed to competent thought, so ignorant that there is knowledge and of what knowledge is, that they do not understand that it matters.

The same terrifying sense of insufficient mental equipment was dawning upon some of us who watched the birth of the League of Nations. Reluctantly and with something like horror, we realised that these people who were, they imagined, turning over a new page and beginning a fresh chapter in human history, knew collectively hardly anything about the formative forces of history. Collectively, I say. Altogether they had a very considerable amount of knowledge, unco-ordinated bits of quite good knowledge, some about this period and some about that, but they had no common understanding whatever of the processes in which they were obliged to mingle and interfere. Possibly all the knowledge and all the directive ideas needed to establish a wise and stable settlement of the world's affairs in 1919 existed in bits and fragments, here and there, but practically nothing had been assembled, practically nothing had been thought out, nothing practically had been done to draw that knowledge and these ideas together into a comprehensive conception of the world. I Put it to you that the Peace Conference at Versailles did not use anything but a very small fraction of the political and economic wisdom that already existed in human brains at that time. And I put it to you as rational creatures that if usage had not chilled our apprehension to this state of affairs, we should regard this as fantastically absurd.

And if I might attempt a sweeping generalisation about the general course of human history in the eighteen years that have followed the War, I believe I should have you with me if I described it as a series of flounderings, violent ill-directed mass-movements, slack here and convulsive action there. We talk about the dignity of history. It is a bookish phrase for which I have the extremest disrespect. There is no dignity yet in human history. It would be pure comedy, if it were not so often tragic, so frequently dismal, generally dishonourable and occasionally quite horrible. And it is so largely tragic because the creature really is intelligent, can feel finely and acutely, expresses itself poignantly in art, music and literature, and—this is what I am driving at—impotently knows better.

Consider only the case of America during this recent period. America when all is said and done, is one of the most intelligently aware communities in the world. Quite a number of people over there seem almost to know what is happening to them. Remember first the phase of fatuous self-sufficiency, the period of unprecedented prosperity, the boom, the crisis, the slump and the dismay. And then appeared the new President, Franklin Roosevelt, and from the point of view of the present discussion he is one of the most interesting figures in all history. Because he really did make an appeal for such knowledge and understanding as existed to come to his aid. America in an astounding state of meekness was ready to be told and shown. There were the universities, great schools, galaxies of authorities, learned men, experts, teachers, gowned, adorned and splendid. Out of this knowledge mass there have since come many very trenchant criticisms of the President's mistakes. But at the time this—what shall I call it—this higher brain, this cerebrum, this grey matter of America was so entirely unco-ordinated that it had nothing really comprehensive, searching, thought-out and trustworthy for him to go upon. The President had to experiment and attempt this and that, he fumed from one promising adviser to another, because there was nothing ready for him. He did not pretend to be a divinity. He was a politician of exceptional good-will. He was none of your dictator gods. He showed himself extremely open and receptive for the organised information and guidance... that wasn't there.

And it isn't there now.

Some years ago there was a considerable fuss in the world about preparedness and unpreparedness. Most of that clamour concerned the possibility of war. But this was a case of a most fantastic unpreparedness on the part of hundreds of eminent men, who were supposed to have studied them, for the normal developments of a community in times of peace. There had been no attempt to assemble that mechanism of knowledge of which America stood in need.

I repeat that if usage had not drilled us into a sort of acquiescence, we should think our species collectively insane to go about its business in this haphazard, planless, negligent fashion.

I think I have said enough to recall to any one here, who may have lapsed from the keen apprehension of his first realisation, this wide gap between what I may call the at present unassembled and unexploited best thought and knowledge in the world, and the ideas and acts not simply of the masses of common people, but of those who direct public affairs, the dictators, the leaders, the politicians, the newspaper directors and the spiritual guides and teachers. We live in a world of unused and misapplied knowledge and skill. That is my ease. Knowledge and thought are ineffective. The human species as a whole is extraordinarily like a man of the best order of brain, who through some lesions or defects or insufficiencies of his lower centres, suffers from the wildest unco-ordinations; St. Vitus's dance, agraphobia, aphonia, and suffers dreadfully (knowing better all the time) from the silly and disastrous gestures he makes and the foolish things he says and does.

I don't think this has ever been so evident as it is now. I doubt if in the past the gap was so wide as it is now between the occasions that confront us, and the knowledge we have assembled to meet them. But because of a certain run of luck in the late nineteenth century, the existence of that widening gap and the menace of that widening gap, was not thrust upon our attention as it has been since the war.

At first that realisation of the ineffectiveness of our best thought and knowledge struck only a few people, like Mr. Maynard Keynes for example, who were in what I may call salient positions, but gradually I have noted the realisation spreading and growing. It takes various forms. Prominent men of science speak more and more frequently of the responsibility of science for the disorder of the world. And if you are familiar with that most admirable of all newspapers, *Nature*, and if you care to turn over the files of that very representative weekly for the past quarter of a century or so and sample the articles, you will observe a very remarkable change of note and scope in what it has to say to its readers. Time was when Nature was almost pedantically special and scientific. Its detachment from politics and general affairs was complete. But latterly the concussions of the social earthquake and the vibration of the guns have become increasingly perceptible in the laboratories. Nature from being specialist has become world—conscious, so that now it is almost haunted week by week by the question: "What are we to do before it is too late, to make what we know and our way of thinking effective in world affairs?" In that I think it is expressing a change which is happening in the minds of—if I may presume to class myself with you—nearly all people of the sort which fills this theatre tonight.

And consider again the topics that have been dealt with at the latest gathering of the British Association. The very title of the Presidential Address: "The Impact of Science upon Society." Sir Josiah Stamp, as you will remember, stressed the need of extending endowment and multiplying workers in the social sciences. Professor Philip dealt with "The Training of the Chemist for the Service of the Community." Professor Cramp talked of "The Engineer and the Nation," and there was an important discussion of "The Cultural and Social Values of Science" in which Sir Richard Gregory, Professor Hogben and Sir Daniel Hall said some memorable things. There can be no doubt of the reality of this awakening of the scientific worker to the necessity of his becoming a definitely organised factor in the social scheme of the years before us.

Well, so far I have been merely opening up my subject and stating the problem for consideration. We want the intellectual worker to become a more definitely organised factor in the human scheme. How is that factor to be organised? Is there any way of implementing knowledge for ready and universal effect? I ask you to examine the question whether this great and growing gap of which we are becoming so acutely aware, between special knowledge and thought and the common ideas and motives of mankind can be bridged, and if so how it can be bridged. Can scientific knowledge and specialised thought be brought into more effective relation to general affairs?

Q:— Let us consider first what is actually going on. I find among my uneasy scientific and specialist friends a certain disposition—and I think it is a mistaken disposition—for direct political action and special political representation. The scientific and literary workers of the days when I was a young man were either indifferent or conservative in politics, nowadays quite a large proportion of them are inclined to active participation in extremist movements; many are leftish and revolutionary, some accept the strange pseudo-scientific dogmas of the Communist party, though that does no credit to their critical training, and even those who are not out on the left are restless for some way of intervening, definitely as a class, in the general happenings of the community. Their ideas of possible action vary from important-looking signed pronouncements and protests to a sort of strike against war, the withholding of services and the refusal I to assist in technical developments that may be misapplied. Some favour the idea of a gradual supersession of the political forms and methods of mass democracy by government through some sort of elite, in which the man of science and the technician will play a dominating part. There are very large vague patches upon this idea, but the general projection is in the form of a sort of modern priesthood, an oligarchy of professors and exceptionally competent people. Like Plato they would make the philosopher king. This project involves certain assumptions about the general quality and superiority of the intellectual worker that I am afraid will not stand scrutiny.

I submit that sort of thing—political activities, party intervention and dreams of an authoritative élite—is not the way in which specialists, artists and specialised thinkers and workers who constitute the vital feeling and understanding of the body politic can be brought into a conscious, effective, guiding and directive relationship to the control of human affairs. Because—I hope you will acquit me of any disrespect for science and) philosophy when I say this—we have to face the fact that from the point of view of general living, men of science, artists, philosophers, specialised intelligences of any sort, do not constitute an elite that can be mobilised for collective action. They are an extraordinarily miscellaneous assembly, and their most remarkable common quality is the quality of concentration in comparative retirement, each along his own line. They have none of the solidarity, the customary savoir faire, the habits arising out of practices, activities and interests in common that lawyers, doctors or any of the really socially organised professions for instance display. A professor-ridden world might prove as unsatisfactory under the stress of modern life and fluctuating conditions as a theologian-ridden world. A distinguished specialist is precious because of his cultivated gift. It does not follow at all that by the standards of all-round necessity he is a superior person. Indeed by the very fact of his specialisation he may be less practised and competent than the average man. He probably does not read his newspaper so earnestly, he finds much of the common round a bother and a distraction and he puts it out of his mind. I think we should get the very gist of this problem if we could compare twelve miscellaneous men of science and special skill, with twelve unspecialised men taken— let us say—from the head clerk's morning train to the city. We should probably find that for commonplace team-work and the ordinary demands and sudden urgencies of life, the 2nd dozen was individually quite as good as, if not better than, the first dozen. In a burning hotel or cast away on a desert island they would probably do quite as well. And yet collectively they would be limited men; the whole dozen of them would have nothing much more to tell you than any one of them. On the other hand our dozen specialists would each have something distinctive to tell you. The former group would be almost as uniform in their knowledge and ability as tiles on a roof; the latter would be like pieces from a complicated jig-saw puzzle. The more you got them together the more they would signify. Twelve clerks or a hundred clerks; it wouldn't matter; you would get nothing but dull repetitions and a flat acquiescent suggestible outlook upon life. But every specialised man we added would be adding something to the directive pattern of life. I think that consideration takes us a step further in defining our problem tonight.

It is science and not men of science that we want to enlighten and animate our politics and rule the world. And now I will take rather a stride forward in my argument. I will introduce a phrase New Encyclopaedism which I shall spend most of the rest of my time defining. I want to suggest that something—a new social organ, a new institution—which for a time I shall call World Encyclopaedia, is the means whereby we can solve the problem of that jig-saw puzzle and bring all the scattered and ineffective mental wealth of our world into something like a common understanding, and into effective reaction upon our vulgar everyday political, social and economic life. I warn you that I am flinging moderation to the winds in the suggestions I am about to put before you. They are immense suggestions. I am sketching what is really a scheme for the reorganisation and reorientation of education and information throughout the world. No less. We are so accustomed to the existing schools, colleges, universities, research organisations of the world; they have so moulded and made us and trained us from our earliest years to respect and believe in them; that it is with a real feeling of temerity, of a matricidal impiety, so to speak, that I have allowed my mind to explore their merits and question whether they are not now altogether an extraordinarily loose, weak and out-of-date miscellany. Yet I do not see how we can admit, and I am disposed to think you have admitted with me, the existence of this terrifying gap between available knowledge and current social and political events, and not go on to something like an indictment of this whole great world of academic erudition, training and instruction from China to Peru—an indictment for, at least, inadequacy and unco- ordination if not for actual negligence. It may be only a temporary inadequacy, a pause in development before renascence, but inadequate altogether they are. Universities have multiplied greatly, yes, but they have failed to participate in the general advance in power, scope and efficiency that has occurred in the past century.

In transport we have progressed from coaches and horses by way of trains to electric traction, motor-cars and aeroplanes. In mental organisation we have simply multiplied our coaches and horses and livery stables. Let me now try to picture for you this missing element in the modern human social mechanism, this needed connection between the percipient and informative parts and the power organisation for which I am using this phrase, World Encyclopaedia. And I will take it first from the point of view of the ordinary educated citizen—for in a completely modernised state every ordinary man will be an educated citizen. I will ask you to imagine how this World Encyclopaedia organisation would enter into his life and how it would affect him. From his point of view the World Encyclopaedia would be a row of volumes in his own home or in some neighbouring house or in a convenient public library or in any school or college, and in this row of volumes he would, without any great toil or difficulty, find in clear understandable language, and kept up to date, the ruling concepts of our social order, the outlines and main particulars in all fields of knowledge, an exact and reasonably detailed picture of our universe, a general history of the world, and if by any chance he wanted to pursue a question into its ultimate detail, a trustworthy and complete system of reference to primary sources of knowledge. In fields where wide varieties of method and opinion existed, he would find, not casual summaries of opinions, hut very carefully chosen and correlated statements and arguments. I do not imagine the major subjects as being dealt with in special articles rather hastily written, in what has been the tradition of Encyclopaedias since the days of Diderot's heroic effort. Our present circumstances are altogether different from his. Nowadays there is an immense literature of statement and explanation scattered through tens of thousands of books, pamphlets and papers, and it is not necessary, it is undesirable, to trust to such hurried summaries as the old tradition was obliged to make for its use. The day when an energetic journalist could gather together a few star contributors and a miscellany of compilers of very uneven quality to scribble him special articles, often tainted with propaganda and advertisement, and call it an Encyclopaedia, is past. The modern World Encyclopaedia should consist of selections, extracts, quotations, very carefully assembled with the approval of outstanding authorities in each subject, carefully collated and edited and critically presented. It would be not a miscellany, but a concentration, a clarification and a synthesis. This World Encyclopaedia would be the mental background of every intelligent man in the world. It would be alive and growing and changing continually under revision, extension and replacement from the original thinkers in the world everywhere. Every university and research institution should be feeding it. Every fresh mind should be brought into contact with its standing editorial organisation. And on the other hand its contents would be the standard source of material for the instructional side of school and college work, for the verification of facts and the testing of statements—everywhere in the world. Even journalists would deign to use it; even newspaper proprietors might be made to respect it.

Such an Encyclopaedia would play the role of an undogmatic Bible to a world culture. It would do just what our scattered and disoriented intellectual organisations of today fall short of doing. It would hold the world together mentally.

It may be objected that this is a Utopian dream. This is something too great to achieve, too good to be true. I won't deal with that for a few minutes. Flying was a Utopian dream a third of a century ago. What I am putting before you is a perfectly sane, sound and practicable proposal.

But first I will notice briefly two objections—obstructions rather than objections—that one will certainly encounter at this point.

One of these is not likely to appear in any great force in this gathering. You have all heard and you have all probably been irritated or bored by the assertion that no two people think alike, *"quot homines, tot sententiae"*, that science is always contradicting itself, that theologians and economists can never agree. It is largely mental laziness on the defensive that makes people say this kind of thing. They don't want their intimate convictions turned over and examined and it is unfortunate that the emphasis put upon minor differences by men of science and belief in their strenuous search for the completest truth and the exactest expression sometimes gives colour to this sort of misunderstanding. But I am inclined to think that most people overrate the apparent differences in the world of opinion today. Even in theology a psychological analysis reduces many flat contradictions to differences in terminology. My impression is that human brains are very much of a pattern, that under the same conditions they react in the same way, and that were it not for tradition, upbringing, accidents of circumstance and particularly of accidental individual obsessions, we should find ourselves— since we all face the same universe—much more in agreement than is superficially apparent. We speak different languages and dialects of thought and can even at times catch ourselves flatly contradicting each other in words while we are doing our utmost to express the same idea. And self-love and personal vanity are not excluded from the intellectual life. How often do we sec men misrepresenting each other in order to exaggerate a difference and secure the gratification of an argumentative victory! A World Encyclopaedia as I conceive it would bring together into close juxtaposition and under critical scrutiny many apparently conflicting systems of statement. It might act not merely as an assembly of fact and statement, but as an organ of adjustment and adjudication, a clearing house of misunderstandings; it would be deliberately a synthesis, and so act as a [test] and a filter for a very great quantity of human misapprehension. It would compel men to come to terms with one another. I think it would relegate *"quot homines, tot sententiae"* back to the Latin comedy from which it emerged.

The second type of obstruction that this idea of a World Encyclopaedia will encounter is even less likely to find many representatives in the present gathering and I will give it only the briefest of attention. (You know that kind of neuralgic expression, the high protesting voice, the fluttering gesture of the hands.) But you want to stereotype people. What a dreadful, dreadful world it will be when everybody thinks alike "—and so they go on. Most of these elegant people who want the world picturesquely at sixes and sevens are hopeless cases, but for the milder instances it may be worth while remarking that it really does not enhance the natural variety and beauty of life to have all the clocks in a town keeping individual times of their own, no charts of the sea, no timetables, but trains starting secretly to unspecified destinations, infectious diseases without notification and postmen calling occasionally when they can get by the picturesque footpads at the corner. I like order in the place of vermin, I prefer a garden to a swamp and the whole various world to a hole-and-corner life in some obscure community, and tonight I like to imagine I am making my appeal to hearers of a kindred disposition to my own. And next let us take this World Encyclopaedia from the point of view of the specialist and the super-intellectual. To him even more than to the common intelligent man World Encyclopaedia is going to be of value because it is going to afford him an intelligible statement of what is being done by workers parallel with himself. And further it will be giving him the general statement of 4 his own subject that is being made to the world at large. He can watch that closely. On the assumption that the World Encyclopaedia is based on a world-wide organisation he will be—if he is a worker of any standing—a corresponding associate of the Encyclopaedia organisation. He will be able to criticise the presentation of his subject, to suggest amendments and re-statements. For a World Encyclopaedia that was kept alive and up to date by the frequent re-issue of its volumes, could be made the basis of much fundamental discussion and controversy. It might breed swarms of pamphlets, and very wholesome swarms. It would give the specialist just that contact with the world at large which at present is merely caricatured by more or less elementary class teaching, amateurish examination work and college administrations. In my dream of a World Encyclopaedia I have a feeling that part of the scheme would be the replacement of the latter group of professional activities, the college business, tutoring, normal lecturing work and so on, by a new set of activities, the encyclopaedic work, the watching brief to prevent the corruption of the popular mind. In enlightening the general mind the specialist will broaden himself. He will be redeemed from oddity, from shy preciousness and practical futility. Well, you begin to see the shape of this project. And you will realise that it is far away from anything like the valiant enterprise of Denis Diderot and his associates a century and a half ago, except in so far as the nature of its reaction upon the world's affairs is concerned. That extraordinary adventure in intellectual synthesis makes this dream credible. That is our chief connection with it. And here I have to make an incidental disavowal. I want to make it clear how little I have to do with what I am now discussing. In order to get some talk going upon this idea of an Encyclopaedia, I have been circulating a short memorandum upon the subject among a number of friends. I did not think to mark it Private, and unhappily one copy seems to have fallen into the hands of one of those minor pests of our time, a personal journalist, who at once rushed into print with the announcement that I was proposing to write a brand new Encyclopaedia, all with my own little hand out of my own little head. At the age of seventy l Once a thing of this sort is started there is no stopping it—and I admit that announcement put me in my place in a pleasantly ridiculous light. But I think after what I have put before you now that you will acquit me of any such colossal ambition. I implore you not to let that touch of personal absurdity belittle the greatness and urgency of the cause I am pleading. This Encyclopaedia I am thinking of is something in which manifestly I have neither the equipment nor the quality to play any but an infinitesimal part. I am asking for it in the role of a common intelligent man who needs it and understands the need for it, both for himself and his world. After that you can leave me out of it. It is just because in the past I have had some experience in the assembling of outlines of knowledge for popular use that I realise, perhaps better than most people, the ineffectiveness of this sort of effort on the part of individuals or small groups. It is something that must be taken up and taken up very seriously—by the universities, the learned societies, the responsible educational organisations if it is to be brought into effective being. It is a super university I am thinking of a world brain; no less. It is nothing in the nature of a supplementary enterprise. I; is a completion necessary to modernise the university. And that brings me to the last part of this speculation. Can such an Encyclopaedia as I have been suggesting to you be a possible thing? How can it be set going? How can it be organised and paid for?

I agree I have now to show it is a possible thing. For I am going to make the large assumption that you think that it is a possible thing it is a desirable thing. How are we to set about it?

I think something in this way: To begin with we want a Promotion Organisation. We want, shall I call it, an Encyclopaedia Society to ask for an Encyclopaedia and get as many people as possible asking for an Encyclopaedia. Directly that Society asks for an Encyclopaedia it will probably have to resort to precautionary measures against any enterprising publisher who may see in that demand a chance for selling some sort of vamped-up miscellany as the thing required, and who may even trust to the unworldliness of learned men for some sort of countenance for his raid.

And next this society of promoters will have to survey the available material. For most of the material for a modern Encyclopaedia exists already—though in a state of impotent diffusion. In all the various departments with which an Encyclopaedia should deal, groups of authoritative men might be induced to prepare a comprehensive list of primary and leading books, articles, statements which taken together would give the best, clearest and most quintessential renderings of what is known and thought within their departments. This would make a sort of key bibliography to the thoughts and knowledge of the world. My friend Sir Richard Gregory has suggested that such a key bibliography for a World Encyclopaedia would in itself. be a worthwhile thing to evoke. I agree with him. I haven't an idea what we should get. I imagine something on the scale of ten or twenty thousand items. I don't know.

Possibly our Encyclopaedia Society would find that such a key bibliography was in itself a not unprofitable publication, but that is a comment by the way. The next step from this key bibliography would be the organisation of a general editorial board and of departmental boards. These would be permanent bodies—for a World Encyclopaedia must have a perennial life. We should have to secure premises, engage a literary staff and, with the constant co-operation of the departmental groups, set about the task of making our great synthesis and abstract. I must repeat that for the purposes of a World Encyclopaedia probably we would not want much original writing. If a thing has been stated clearly and compactly once for all, why paraphrase it or ask some inferior hand to restate it? Our job may be rather to secure the use of copyrights, and induce leading exponents of this or that field of science or criticism to co-operate in the selection, condensation, expansion or simplification of what they have already said so well.

And now I will ask you to take another step forward and imagine our World Encyclopaedia has been assembled and digested and that the first edition is through the press. So far we shall have been spending money on this great enterprise and receiving nothing; we shall have been spending capital, for which I have at present not accounted. I will merely say that I see no reason why the capital needed for these promotion activities should not be forthcoming. This is no gainful enterprise, but you have to remember that the values we should create would be far more stable than the ephemeral encyclopaedias representing sums round about a million pounds or so which have hitherto been the high-water of Encyclopaedic enterprise. These were essentially book-selling enterprises made to exploit a demand. But this World Encyclopaedia as I conceive it, if only because it will have roped in the larger part of the original sources of exposition, discussion and information, will be in effect a world monopoly, and it will be able to levy and distribute direct and indirect revenue, on a scale quite beyond the resources of any private publishing enterprise. I do not see that the financial aspects of this huge enterprise, big though the sums involved may be, present any insurmountable difficulties in the way of its realisation. The major difficulty will be to persuade the extremely various preoccupied, impatient and individualistic scholars, thinkers, scientific workers and merely distinguished but unavoidable men on whose participation its success depends, of its practicability, convenience and desirability. And so far as the promotion of it goes I am reasonably hopeful. Quite a few convinced, energetic and resourceful people could set this ball rolling towards realisation. To begin with it is not necessary to convert the whole world of learning, research and teaching. I see no reason why at any stage it should encounter such positive opposition. Negative opposition—the refusal to have anything to do with it and so forth-can be worn down by persistence and the gathering promise of success. It has not to fight adversaries or win majorities before it gets going. And once this ball is fairly set rolling it will be very hard to stop. A greater danger, as I have already suggested, will come from attempts at the private mercenary exploitation of this world-wide need—the raids of popular publishers and heavily financed salesmen, and in particular attempts to create copyright difficulties and so to corner the services and prestige of this or that unwary eminent person by anticipatory agreements.

Vis-à-vis with salesmanship the man of science, the man of the intellectual élite, is a t to show himself a very Simple Simon indeed. And) of course from the very start, various opinionated cults and propagandists will be doing their best to capture or buy the movement. Well, we mustn't be captured or bought, and in particular our silence must not be bought or captured. That danger may in the end prove to be a stimulus. It may be possible in some cases to digest and assimilate special cults to their own and the general advantage.

And there will be a constant danger that some of the early promoters may feel and attempt to realise a sort of proprietorship in the organisation, to make a group or a gang of it. But to recognise that danger is half-way to averting it.

I have said nothing so far about the language in which the Encyclopaedia should appear. It is a question I have not worked out. But I think that the main text should be in one single language, from which translations in whole or part could be made. Catholic Christianity during the years of its greatest influence was held together by Latin, and I do not think I am giving way to any patriotic bias when I suggest that unless we contemplate a polyglot publication—and never yet have I heard of a successful polyglot publication—English because it has a wider range than German, a greater abundance and greater subtlety of expression than French and more precision than Russian, is the language in which the original text of a World Encyclopaedia ought to stand. And moreover it is in the English-speaking communities that such an enterprise as this is likely to find the broadest basis for operations, the frankest criticism and the greatest freedom from oficial interference and government propaganda. But that must not hinder us from drawing help and contributions from, and contemplating a use in every community in the world.

And so far I have laid no stress upon the immense advantage this enterprise would have in its detachment from immediate politics, Ultimately if our dream is realised it must exert a very great influence upon everyone who controls administrations, makes wars, directs mass behaviour, feeds, moves, starves and kills populations. But it does not immediately challenge these active people. It is not the sort of thing to which they would be directly antagonistic. It is not ostensibly anti-them. It would have a terrible and ultimately destructive aloofness. They would not easily realise its significance for all that they do and are. The prowling beast will right savagely if it is pursued and challenged upon the jungle path in the darkness, but it goes home automatically as the day breaks.

You see how such an Encyclopaedic organisation could spread like a nervous network, a system of mental control about the globe, knitting all the intellectual workers of the world through a common interest and a common medium of expression into a more and more conscious co-operating unity and a growing sense of their own dignity, informing without pressure or propaganda, directing without tyranny. It could be developed wherever conditions were favourable; it could make inessential concessions and bide its time in regions of exceptional violence, grow vigorously again with every return to liberalism and reason.

So I sketch my suggestion for a rehabilitation of thought and learning that ultimately may release a new form of power in the world, recalling indeed the power and influence of the churches and religions of the past but with a progressive, adaptable and recuperative quality that none of these possessed. I believe that in some such way as I have sketched tonight the mental forces now largely and regrettably scattered and immobilised in the universities, the learned societies, research institutions and technical workers of the world could be drawn together in a real directive world intelligence, and by that mere linking and implementing of what is known, human life as a whole could be made much surer, stronger, bolder and happier than it has ever been up to the present time. And until something of this sort is done, I do not see how the common life can ever be raised except occasionally, locally and by a conspiracy of happy chances, above its present level of impulsiveness, insincerity, insecurity, general under-vitality, under-nourishment and aimlessness. For that reason I think the promotion of an organisation for a World Encyclopaedia may prove in the long run to be a better investment for the time and energy of intelligent men and women than any definite revolutionary movement, Socialism, Communism, Fascism, Imperialism, Pacifism or any other of the current isms into which we pour ourselves and our resources so freely. None of these movements have anything like the intellectual comprehensiveness needed to construct the world anew.

Let me be very clear upon one point.

I am not saying that a World Encyclopaedia will in itself solve any single one of the vast problems that must be solved if man is to escape from his present dangers and distresses and enter upon a more hopeful phase of history; what I am saying—and saying with the utmost conviction—is this, that without a World Encyclopaedia to hold men's minds together in something like a common interpretation of reality, there is no hope whatever of anything but an accidental and transitory alleviation of any of our world troubles. As mankind is, so it will I remain, until it pulls its mind together. And if it does . not pull its mind together then I do not see how it can help but decline. Never was a living species more perilously poised than ours at the present time. If it does not take thought to end its present mental indecisiveness catastrophe lies ahead. Our species may yet end its strange eventful history as just the last, the cleverest of the great apes.

The great ape that was clever—but not clever enough. It could escape from most things but not from its own mental confusion.

II. THE BRAIN ORGANIZATION
OF THE MODERN WORLD

LECTURE DELIVERED IN AMERICA,
OCTOBER AND NOVEMBER, 1937

FOR half a century I have resisted temptations to lecture in America—if for no other reason than the insufficiency of my voice. But the microphone is a great leveller and here I am at last on terms of practical equality with your most audible speakers and very glad indeed of this belated opportunity of talking to you. I want to talk to you about an idea which seems to me to be a very important one indeed. I want to interest you in it, and if possible find out what you think of it. I call that idea for reasons I shall try to make clear as I proceed, The New Encyclopaedism, and the gist of it is that the time is ripe for a very extensive revision and modernisation of the intellectual organisation of the world. Can I put it more plainly than that? Perhaps I can.

Our world is changing and it is changing with an ever-increasing violence. An old world dies about us. A new world struggles into existence. But it is not developing the brain and the sensitiveness and delicacy necessary for its new life. That is the essence of what I have to say.

To put my argument squarely on its feet I must begin by telling you things that you know quite as well or better than I do. I will just remind you of them. It is, so to speak, a matter of current observation that in the past century and a half there has been an enormous increase in the speed and facility of communications between men in every part of the world. Two hundred years ago Oliver Goldsmith said that if every time a man fired a gun in England, someone was killed in China, we should never hear of it and no one would bother very much about it. All that is changed. We should hear about that murdered Chinaman almost at once. Today we can go all round the world in the time it took a man to travel from New York to Washington in 1800, we can speak to any one anywhere so soon as the proper connections have been made and in a little while we shall be able to look one another in the face from the ends of the earth. In a very few years now we shall be able to fly in the stratosphere across the Atlantic in a few hours with a cargo of passengers, or bombs or other commodities. There has in fact been a complete revolution in our relation to distances. And the practical consequences of these immense approximations are only beginning to be realised. Everybody knows these facts now, but round about 1900 we were only beginning to take notice of this abolition of distance. Even in 1919 the good gentlemen who settled the world for ever at Versailles had not observed this strange new thing in human affairs. They had not observed that it was no longer possible to live in little horse-and-foot communities because of this change of scale. We know better now. Now the consequences of this change of scale force themselves upon our attention everywhere. Often in the rudest fashion. Our interests and our activities interpenetrate more and more. We are all consciously or unconsciously adapting ourselves to a single common world. For a time North America and the great sprawl of Russia and

Siberia are for obvious reasons feeling less restriction than let us say japan or Germany, but, as my glancing allusion to the stratosphere was intended to remind you, this relative isolation of yours is also a diminishing isolation. The Abolition of Distance is making novel political and economic arrangements more and more imperative if the populations of the earth are not to grind against each other to their mutual destruction.

That imperative expansion of the scale of the community in which we have to live is the first truism I want to recall to you and bring into the foreground of our discussion. The second truism is the immense increase in our available power that has been going on. I do not know if any precise estimate of the physical energy at the disposal of mankind now and at any previous age, has ever been made, but the disproportion between what we have and what our great-grand-parents had, is stupendous and continually increasing. I am told that two or three power stations in the United States are today pouring out more energy night and day than could be produced by the sustained muscular effort of the entire United States population, and that the Roman empire at its mightiest could not—even by one vast unanimous thrust, not a single soul doing anything but push and push—have kept the street and road transport of New York State moving as it moves today. You are almost sick of being told it, in this form or that, over and over again. But we all know about this sort of thing. Man was slower and feebler beyond comparison a century or so ago than he is today. He has become a new animal incredibly swift and strong—except in his head. We all know—in theory at least—how this increase of power affects the nature of war. None of our new powers in this world of increasing power, have been so rapidly applied as our powers of mutual injury. A child of five with a bomb no bigger than my hand, can kill as many men in a moment as any paladin of antiquity hacking and hewing and bashing through a long and tiring battle. Both these two realities, these two portentous realities, the change of scale in human affairs and the monstrous increase of destructive power, haunt every intelligent mind today. One needs an exceptional stupidity even to question the urgency we are under to establish some effective World Pax, before gathering disaster overwhelms us. The problem of reshaping human affairs on a world-scale, this World problem, is drawing together an ever-increasing multitude of minds. It is becoming the common solicitude of all sane and civilised men. We must do it—or knock ourselves to pieces. I think it would be profitable if a group of history students were to trace how this World Problem has dawned upon the popular mind from, let us say, 1900 up to the present time. To begin with it was hardly felt to be important. Our apprehension of what it really amounts to has grown in breadth and subtlety during all these past seven-and-thirty years. We have been learning hard in the past third of a century. And particularly since 1919. In 1900 the general sense of the historical process, of what was going on in the world, was altogether shallower than ours today. People were extraordinarily ignorant of the operating causes of political events. It was quite possible then for them to agree that it was not at all a nice or desirable thing I and that it ought to be put an end to, and to imagine that setting up a nice little international court at the Hague to which states could bring their grievances and get a decision without going to the trouble and expense of hostilities would end this obsolescent scandal. Then we should have peace for ever—and everything else would go on as before. But now even the boy picking cotton or working the elevator, knows that nothing will go as before. The fear of change has reached them. You will remember that Mr. Andrew Carnegie set aside quite a respectable fraction of his savings to buy us world peace for ever and have done with it. The Great War was an enlightening disappointment to this earlier school of peacemakers, and it released a relatively immense flow of thought about the World Problem. But even at Versailles the people most immediately powerful, were still evidently under the impression that world peace was simply a legal and political business. They thought the Great War had happened, but they were busy politicians, and had not remarked that vastly greater things were happening. They did not realise even that elementary point about the unsuitable size of contemporary states to which I have recalled your attention. Still less did they think about the new economic stresses that were revolutionising every material circumstance of litre. They saw the issue as a simple aiirair upon the lines of old-fashioned history. So far as their ideas went it was just Carthage and Rome over again. The central Powers were naughty naughty nations and had to be punished. Their greatest novelty was the League of Nations, which indeed was all very well as a gesture and an experiment but which as an irremovable and irreplaceable reality in the path of world adjustment has proved anything but a blessing. It had been a brilliant idea in the reign of Francis I of France. Still we have to recognise that in 1919 the Geneva League was about as far as anyone's realisation of the gravity of the World Problem had gone. It is our common quality to be wise after the event and still quite unprepared for the next change ahead. It is an almost universal human failing to believe that now we know everything, that nothing more than we know can be known about human relations, and that in our limitless wisdom we can fix up our descendants for evermore, by constitutions, treaties, boundaries and leagues. So my poor generation built this insufficient League. For a time a number of well-meaning people did consider that the League of Nations settled the World Problem for good and all, and that they need not bother their heads about it any more. There were we felt, no further grounds for anxiety, and we all sat down within our nice little national boundaries to resume business . according to the old ways, securing each of us the largest possible share of the good things the new Era of Peace and Prosperity was to bring—at least to the good countries to whom victory had been accorded. Wlten later the history of our own times comes to be written, I imagine this period between 1919 and 1929 will be called the Fatuous Twenties.

We all know better now. Now that we are living in what no doubt the historian will some day call the Frightened Thirties. Versailles was no settlement. There is still no settlement. The World Problem still pursues us. And it seems now vastly nearer, uglier and more formidable than it ever did before. It emerges through all our settlements like a dangerous rhinoceros coming through a reed fence. Our mood changes now from one in which off-hand legal solutions were acceptable, to an almost feverish abundance of mental activity. From saying "There is the Hague Court and what more do you want?" or "There is the League of Nations, what more can you want?" or "There is the British Peace Ballot and please don't bother me further," we are beginning to apprehend something of the full complexity and vastness of the situation that faces mankind, that is to say all of us, as a living species. Our minds are beginning to grasp the vastness of these grim imperatives. That change of scale, that enhancement of power has altered the fundamental conditions of human life—of all our lives. The traditions of the old world, the comparatively easy traditions in which we have grown up and in which we have shaped our lives, are bankrupt. They are outworn. They are outgrown. They are too decayed for much more patching. They are as untrustworthy and dangerous as a very old car whose engine has become explosive, which has lost its brake lining and has a loose steering-wheel. What I am saying now is gradually becoming as plain in men's minds as the roundness of the earth. New World or nothing. We have to make a new world for ourselves or we shall suffer and perish amidst the downfall of the decaying old. This is a business of fundamentals in which we are all called upon to take part, and through which the lives of all of us are bound to be changed essentially and irrevocably.

With this realisation of the true immensity and penetration of the World Problem we are passing out of the period of panaceas—of simple solutions. As We grow wiser we realise more and more that the World Problem is not a thing like a locked door for which it is only necessary to find a single key. It is infinitely more complex. It is a battle all along the line and every man is a combatant or a deserter. Popular discussion is thick with competing simple remedies, these [sometimes] needful proposals, each of which has its factor of truth and each of which in itself is entirely inadequate. Consider some of them. Arbitration, League of Nations, I have spoken of World Socialism. The Socialist very rightly points out the evils and destructive stresses that arise from the free play of the acquisitive impulse in production and business affairs, but his solution, which is to take the control of things out of the hands of the acquisitive in order to put it into the hands of the inexperienced, plainly leaves the bulk of the world's troubles unsolved. The Communist and Fascist have theorised about and experimented with the seizure and concentration of Power, but they produce no sound schemes for its beneficial use. Seizing power by itself is a gangster's game. You can do nothing with power except plunder and destroy—unless you know exactly what to do with it. People tell us that Christianity, the Spirit of Christianity, holds a key to all our difficulties. Christianity, they say, has never yet been tried. We have all heard that. The trouble is that Christianity in all its various forms never does try. Ask it to work out practical problems and it immediately floats off into other-worldliness. Plainly there is much that is wrong in our property-money arrangements, but there again prescriptions for a certain juggling with currency and credit, seem unlikely in themselves to solve the World Problem. A multitude of such suggestions are bandied about with increasing passion. In comparison with any preceding age, we are in a state of extreme mental fermentation. This is, I suggest, an inevitable phase in the development of our apprehension of the real magnitude and complexity of the World Problem which faces us. Except for the faddists and fanatics we all feel a sort of despairing inadequacy amidst this wild storm of suggestions and rash beginnings. We want to know more, we want digested facts to go upon. Our minds are not equipped for the job.

And shaking a finger at you to mark the point we have reached, I repeat, our minds are not equipped for the job.

We are ships in uncharted seas. We are big-game hunters without weapons of precision.

This present uproar of incomplete ideas was as inevitable as the Imperialist Optimism of 1900, the Futile Amazement of the

Great War, and the self-complacency of the Fatuous Twenties. These were all phases, necessary phases, in the march of our race through disillusionment to understanding. After the phase of panaceas there comes now, I hope, a phase of intelligent co-ordination of creative movements, a balanced treatment of our complex difficulties. We are going to think again. We are all beginning to realise that the World Problem, the universal world problem of adapting our life to its new scale and its new powers, has to be approached on a broad front, along many paths and in many fashions. In my opening remarks I stressed our spreading realisation of the possibility of a great catastrophe in world affairs. One immediate consequence of our full realisation of what this World Problem before us means is dismay. We lose heart. We feel that anyhow we cannot adjust that much. We throw up the sponge. We say, let us go on as long as possible anyhow, and after us, let what will happen. A considerable and a growing number of people are persuaded that a drift towards a monstrously destructive war cycle which may practically obliterate our present civilisation is inevitable. I have, I suppose, puzzled over such possibilities rather more than most people. I do not agree with that inevitability of another real war. But I agree with its possibility. I think such a collapse so possible that I have played with it imaginatively in a book or so and a film. It is so much a possibility that it is wholesome to bear it constantly in mind. But all the same I do not believe that world disaster is unavoidable. It is extraordinarily difficult to estimate the relative strength of the driving forces in human affairs today. We are not dealing with measurable quantities. We are easily the prey of our moods, and our latest vivid impression is sure to count for far too much. Values in my own mind, I find, shift about from hour to hour. I guess it is about the same with most of you. Just as in a battle, so here, our moods are factors in the situation. When we feel depressed, the world is going to the devil and we meet defeat half-way; when we are elated, the world is all right and we win. And I think that most of us are inclined to overestimate the menace of violence, the threats of nationalist aggression and the suppression of free discussion in many parts of the world at the present time. I admit the darkness and grimness on the face of things. Indisputably vehement State-ism now dominates affairs over large regions of the civilised world. Everywhere liberty is threatened or outraged. Here again, I merely repeat, what the whole intelligent world is saying.

Well....I do not want to seem smug amidst such immunities as we English- speaking people still enjoy, nevertheless I must confess I think it possible to overrate the intensity and staying power of this present nationalist phase. I think that the present vehemence of nationalism in the world may be due not to the strength of these tyrannies but to their weakness. This change of scale, this increment of power that has come into human affairs, has strained every boundary, every institution and every tradition in the world. It is an age of confusion, an age of gangster opportunity. After the gangsters the Vigilantes. Both the dying old and the vamped-up new are on the defensive. They build up their barriers and increase their repression because they feel the broad flood of change towards a vastly greater new order is rising. Every old government, every hasty new government that has leapt into power, is made crazy by the threat of a wider and greater order, and its struggle to survive becomes desperate. It tries still to carry onto deny that it is an experiment—even if it survives crippled and monstrous. The dogmatic Russian Revolution has not held power for a score of years and yet it, too, is now as much on the defensive as any other upstart dictatorship. A lot of what looks to us now like triumphant reaction may in the end prove to be no more than doomed, dwarfed and decaying dogmas and traditions at bay. None of the utterances of these militant figures that most threaten the peace of the world today have the serene assurance of men conscious that they are creating something that marches with the ruling forces of life. For the most part they are shouts—screams—of defiance. They scold and rant and threaten. That is the rebel note and not the note of mastery. We hear very much about the suppression of thought in the world. is there really—even at the present time—in spite of all this current violence, any real diminution of creative thought in the world— as compared with 1500 or 1850—or 1900, or 1914 or 1924? You have to remember that the suppression of free discussion in such countries as Germany, Italy and Russia does not mean an end to original thought in these countries. Thought, like gunpowder, may be all the more effective for being confined. I know that beneath the surface Germany is thinking intensely, and Russia is thinking more clearly if less discursively than ever before. Maybe we overestimate the value of that idle and safe, slack, do-as-you-please discussion that we English-speaking folk enjoy under our democratic regime. The concentration camps of today may prove after all to be the austere training grounds of a new freedom.

Let us glance for a moment at the chief forces that are driving against all that would keep the world in its ancient tradition of small national governments, warring and planning perpetually against each other, of a perpetual struggle not only of nations but individuals for a mere cramped possessiveness.

Consider now the drives towards release, abundance, one

World Pax, one world control of violence, that are going on today. They seem to me very much like those forces that drove the United States to the Pacific coast and prevented the break-up of the Union. No doubt, many a heart failed in the covered waggons as they toiled westward, face to face with the Red Indian and every sort of lawless violence. Yet the drive persisted and prevailed. The Vigilantes prepared the way for the reign of law. The railway, the telegraph and so on followed the covered waggon and knitted this new-scale community of America together. In the middle nineteenth century all Europe thought that the United States must break up into a lawless confusion. The railway, the printing press, saved that. The greater unity conquered because of its immense appeal to common-sense in the face of the new conditions. And because it was able to appeal to common sense through these media.

The United States could spread gigantically and keep a common mind. And today I believe in many ways, in a variety of fashions and using many weapons and devices, the Vigilantes of World Peace, under the stimulus of still wider necessities, are finding themselves and each other and getting together to ride. That is to say their minds are getting together. One great line of development must be towards a Common Control of the Air. The great spans of the Atlantic and Pacific may prevent this from beginning as a world-wide Air Control, but that, I think, is just a passing phase of the problem. I submit to you that a state of affairs in which vast populations are under an ever- increasing threat of aerial bombardment with explosives, incendiary bombs and poison gas at barely an hour's notice, is intolerable to human reason. Maybe there will be terrible wars first. Quite possibly not. It may after all prove unnecessary to have very many great cities destroyed and very many millions of people burnt, suffocated, blown limb from limb, before men see what stares them in the face and accept the obvious. Men are, after all, partly reasonable creatures, they have at least spasmodic moral impulses. There is already in action—a movement—for World Air Control. But you can't have a thing like that by itself. Who or what will control the air?

This is a political question. None of us quite know the answer, but the answer has to be found, and hundreds of thousands of the best brains on earth are busy at the riddle of that adjustment. We can rule out any of the pat, ready-made answers of yesterday, League of Nations or what not. None the less that implacable necessity for World Air Control insists upon something, something with at least the authority of a World Federal Government in these matters, and that trails with it, you will find, a revelation of other vast collateral necessities. I cannot now develop these at any great length. But in the end I believe we are led to the conviction that the elemental forces of human progress, the stars in their courses, are fighting to evoke at least this much world community as involves a control of communications throughout the whole world, a common federal protection of everyone in the world from private, sectarian or national violence, a common federal protection of the natural resources of the planet from national, class or individual appropriation, and a world system of money and credit. The obstinacy of man is great, but the forces that grip him are greater and in the end, after I know not what wars, struggles and afflictions, this is the road along which he will go. He has to see it first—and then he will do it. I am sure of the ultimate necessity of this federal world state—and at the backs of your minds at least, I believe most of you are too—as I am sure that, whatever clouds may obscure it, the sun will rise to-morrow.

And now having recapitulated and brought together this general conception of human progress towards unity which is forming in most of our minds, as an answer to the ever-more insistent World Problem, I propose to devote the rest of my time with you to the discussion of one particular aspect of this march towards a world community, the necessity it brings with it, for a correlated educational expansion. This has not so far been given anything like the attention it may demand in the near future. We have been gradually brought to the pitch of imagining and framing our preliminary ideas of a federal world control of such things as communications, health, money, economic adjustments, and the suppression of crime. In all these material things we have begun to foresee the possibility of a world-wide network being woven between all men about the earth. So much of the World Peace has been brought into the range of—what shall I call it?—the general imagination. But I do not think we have yet given sufficient attention to the prior necessity, of linking together its mental organisations into a much closer accord than obtains at the present time. All these ideas of unifying mankind's affairs depend ultimately for their realisation on mankind having a unified mind for the job. The want of such effective mental unification is the key to most of our present frustrations. While men's minds are still confused, their social and political relations will remain in confusion, however great the forces that are grinding them against each other and however tragic and monstrous the consequences.

Now I know of no general history of human education and discussion in existence. We have nowadays—in what is called the New History-books which trace for us in rough outline the growth in size and complexity of organised human communities. But so far no one has attempted to trace the stages through which teaching has developed, how schools began, how discussions grew, how knowledge was acquired and spread, how the human intelligence kept pace with its broadening responsibilities. We know that in the small tribal community and even in the city states of, for example, Greece, there was hardly any need for reading or writing. The youngsters were instructed and initiated by their elders. They could walk all over the small territory of their community and see and hear, how it was fed, guarded, governed. The bright young men gathered for oral instruction in the Porch or the Academy. With the growth of communities into states and kingdoms we know that the medicine man was replaced by an organised priesthood, we know that scribes appeared, written records. There must have been schools for the priests and scribes, but we know very little about it. We know something of the effect of the early writings, the Bible particularly, in consolidating and preserving the Jewish tradition—giving it such a start off that for a long time it dominated the subsequent development of the Gentile world, and we know that the survival and spread of Christianity is largely due to its resort to written records to supplement that oral teaching of disciples with which it began. But the growing thirst for medical, theological and general knowledge that appeared in the Middle Ages and which led to those remarkable gatherings of hungry minds, the Universities, has still to be explained and described. That appearance and that swarming of scholars would make an extraordinary story. After the lecture room, the book; after that the newspaper, universal education, the cinema, the radio. No one has yet appeared to make an orderly, story of the developments of information and instruction that have occurred in the past hundred years. Age by age the World's Knowledge Apparatus has grown up. Unpremeditated. Without a plan. But enlarging due possible areas of political co-operation at every stage in its growth.

It is a very interesting thing indeed to ask oneself certain questions. How did I come to know what I know about the world and myself? What ought I to know? What would I like to know that I don't know? If I want to know about this or that, where can I get the clearest, best and latest information? And where did these other people about me get their ideas about things?

Which are sometimes so different from mine. Why do we differ so widely? Surely about a great number of things upon which we differ there is in existence exact knowledge? So that we ought not to differ in these things. This is true not merely about small matters in dispute but about vitally important things concerning our business, our money, our political outlook, our health, the general conduct of our lives. We are guessing when we might know. The facts are there, but we don't know them completely. We are inadequately informed. We blunder about in our ignorance and this great ruthless world in which we live, beats upon us and punishes our ignorance like a sin. Not only in our mass-ruled democracies but in the countries where dogmas and dictators rule, tremendous decisions are constantly being made affecting human happiness, root and branch, in complete disregard of realities that are known.

You see we are beginning to realise not only that the formal political structures of the world and many of the methods of our economic life are out of date and out of scale, but also another thing that hampers us hopelessly in every endeavour we make to adjust life to its new conditions—our World Knowledge Apparatus is not up to our necessities. We are neither collecting, arranging nor digesting what knowledge we have at all adequately, and our schools, our instruments of distribution are old-fashioned and ineffective.

We are not being told enough, we are not being told properly, and that is one main reason why we are all at sixes and sevens in our collective life.

The other day my university, the University of London, celebrated its centenary. For some minor reason I was asked to assist at these celebrations. And to do so I had to assume some very remarkable garments—most remarkable if you consider that London University was founded in the year 1836 when gentlemen wore tight trousers with straps, elegantly waisted coats and bell-shaped top hats. Did I dress up like that? No. I found myself retreating from the age of the aeroplane to the age of the horse and mule outfit of the Canterbury Pilgrims. I found myself wearing a hood and gown and carrying a Beret rather like those worn by prosperous citizens of the days of Edward IV, when the University of London was as little anticipated as the continent of America. My modern head peeped out at the top of this get-up and my modern trousers at the bottom. Properly I ought to have been wearing a square beard or have been clean-shaven, but I was forgiven that much. And from all parts of the world representatives of innumerable universities had come with beautifully illuminated addresses to congratulate our Chancellor and ourselves on our hundred years of sham mediaevalism. They came from the ends of the earth, they came up the aisle in an endless process; one ancient name followed another, now it was Tokyo, now Athens, now Upsala, now Cape Town, now the

Sorbonne, now Glasgow, now Johns Hopkins, on they came and on and bowed and handed their addresses and passed aside. It was a marvellous, a dazzling array of beautifully coloured robes. It was also a marvellous collection of men and women. I watched the grave and dignified faces of some of the finest minds in the world. Together they presented, they embodied or they were there to represent, the whole body of human knowledge. There it was in effect parading before me. And nine out of ten of them were dressed up in some colourful imitation of a costume worn centuries before their foundations came into existence. It was picturesque, it was imposing—but it was just a little odd of them.

My thoughts drifted away to certain political gatherings I had seen and heard; faces of an altogether inferior type, leather-lunged adventurers bawling and gesticulating, raucous little men screaming plausible nonsense to ignorant crowds, supporters herded like sheep and saluting like trained monkeys, and the incongruity of the contrast came to me—you know how things come to you suddenly at times—so that I almost laughed aloud. Because, when it comes to the direction of human affairs, all these universities, all these nice refined people in their lovely gowns, all this visible body of human knowledge and wisdom, has far less influence upon the conduct of human affairs, than, let us say, an intractable newspaper proprietor, an unscrupulous group of financiers or the leader of a recalcitrant minority.

Some weeks previously I had taken part in a little private conference of scientific men in London. They were very distinguished men indeed, and they were distressed beyond measure at the way in which one scientific invention after another was turned to the injury of human life. What was to be done? What could be done? Our discussion was inconclusive, but it had quickened my sense of the reality of the situation. I put these three separate impressions together before you: First, these anxious scientific specialists, then the unchallenged power and mischief of these bawling war-making politicians and their crowds at the present time, and finally, capping the whole, these hundreds of all-too-decorated learned gentlemen, fine and delicate, bowing, presenting addresses (for the most part in Latin) and conferring further gowns and diplomas on one another. This last lot, I said, this third lot is after all—in spite of its elegant weakness, the organised brain of mankind so far as there is an organised brain of mankind—and it is not doing its proper work. Why? Why are our universities Boating above the general disorder of mankind like a beautiful sunset over a battlefield? Is it not high time that something was done about it? Certain ideas had been stirring in my mind for some time already, but this scene of archaic ceremony just lit up the situation for me. I realised that these mediaeval robes were in the highest degree symptomatic. They clothed an organisation essentially mediaeval, inadequate and out of date. We are living in 1937 and our universities, I suggest, are not half-way out of the fifteenth century. We have made hardly any changes in our conception of university organisation, education, graduation, for a century—for several centuries. The three- or four-years' course of lectures, the bachelor who knows some, the master who knows most, the doctor who knows all, are ideas that have come down unimpaired from the Middle Ages. Nowadays no one should end his learning while he lives and these university degrees are preposterous. It is true that we have multiplied universities greatly in the past hundred years, but we seem to have multiplied them altogether too much upon the old pattern.

A new battleship, a new aeroplane, a new radio receiver is always an improvement upon its predecessor. But a new university is just another imitation of all the old universities that have ever been. Educationally we are still for all practical purposes in the coach and horse and galley stage. The new university is just one more mental gilt-coach in which minds take a short ride and get out again. We have done nothing to co-ordinate the work of our universities in the world—or at least we have done very little. What are called the learned societies with correspondents all over the world have been the chief addition to the human knowledge organisation since the Renaissance and most of these societies took their shape and scale in the eighteenth and nineteenth centuries. All the new means of communicating ideas and demonstrating realities that modern invention has given us, have been seised upon by other hands and used for other purposes; these universities which should guide the thought of the world, making no protest. The showmen got the cinema and the governments or the adventurers got the radio. The university teacher and the schoolmaster went on teaching in the class-room and checking his results by a written examination. It is as if one attempted to satisfy the traffic needs of greater New York or London or Western Europe by a monstrous increase in horses and carts and nothing else. The universities go out to meet the tremendous challenges of our social and political life, like men who go out in armour with bows and arrows to meet a bombing aeroplane. They are pushed aside by men like Hitler, Mussolini creates academies in their despite, Stalin sends party commissars to regulate their researches. It is beyond dispute that there has been a great increase in the research work of universities; that pedantry and mere scholarship in spite of an obstinate defence have declined relatively to keen inquiry, but the specialist is by his nature a preoccupied man. He can increase knowledge, but without a modern organisation backing him he cannot put it over. He can increase knowledge which ultimately is power, but he cannot at the same time control and spread this power that he creates. It has to be made generally available if it is not to be monopolised in the wrong hands.

There, I take it, is the gist of the problem of World Knowledge that has to be solved. A great new world is struggling into existence. But its struggle remains catastrophic until it can produce an adequate knowledge organisation. It is a giant birth and it is mentally defective and blind. An immense and ever-increasing wealth of knowledge is scattered about the world today, a wealth of knowledge and suggestion that—systematically ordered and generally disseminated—would probably give this giant vision and direction and suffice to solve all the mighty difficulties of our age, but that knowledge is still dispersed, unorganised, impotent in the face of adventurous violence and mass excitement. In some way we want to modernise our World Knowledge Apparatus so that it may really bring what is thought and known within reach of all active and intelligent men. So that we shall know—with some certainty. So that l we shall not be all at sixes and sevens about matters that have already been thoroughly explored and worked out.

How is that likely to he done?

Not of course in a hurry....

It would be very easy to do a number of stupid things about it—futile or even disastrous things. I can imagine quite a number of obvious preposterous mischievous experiments, a terrible sort of world university consolidation, an improvised knowledge dictatorship. Heaven save us from that! We Want nothing that will in any sense override the autonomy of institutions or the independence of individual intellectual workers, We want nothing that will invade the precious time and attempt to control the resources of the gifted individual specialist. He is too much distracted by elementary teaching and college administration already. We do not want to magnify and stereotype universities. Most of them with their gowns and degrees, their slavish imitation of the past, are too stereotyped already. But here it is that the idea I Want to put before you comes in, this idea of a greater encyclopaedism—with a permanent organism and a definite form and aim. I put forward the development of this new encyclopaedism as a possible method, the only possible method I can imagine, of bringing the universities and research institutions of the world into effective co-operation and creating an intellectual authority sufficient to control and direct our collective life. I imagine it as a permanent institution— untrammelled by precedent, a new institution—something added to the world network of universities, linking and co-ordinating them with one another and with the general intelligence of the world. Manifestly as my title for it shows, it arises out of the experience of the French Encyclopaedia, but the form it is taking in the minds of those who have become interested in the idea, is of something vastly more elaborate, more institutional and far-reaching than Diderot's row of volumes. The immense effect of Diderot's effort in establishing the frame of the progressive world of the nineteenth century, is certainly the inspiration of this new idea. The great role played in stabilising and equipping the general intelligence of the nineteenth-century world by the French, the British and the German and other encyclopaedias that followed it, is what gives confidence and substance to this new conception. But what we want today to hold the modern mind together in common sanity is something far greater and infinitely more substantial than those earlier encyclopaedias. They served their purpose at the time, but they are not equal to our current needs. A World Encyclopaedia no longer presents itself to a modern imagination as a row of volumes printed and published once for all, but as a sort of mental clearing house for the mind, a depot where knowledge and ideas are received, sorted, summarised, digested, clarified and compared. It would be in continual correspondence with every university, every research institution, every competent discussion, every survey, every statistical bureau in the world. It would develop a directorate and a staff of men of its own type, specialised editors and summarists. They would be very important and distinguished men in the new world. This Encyclopaedic organisation need not be concentrated now in one place; it might have the form of a network. It would centralise mentally but perhaps not physically. Quite possibly it might to a large extent be duplicated. It is its files and its conference rooms which would be the core of its being, the essential Encyclopaedia. It would constitute the material beginning of a real World Brain.

Then from this centre of reception and assembly, would proceed what we may call the Standard Encyclopaedia, the primary distributing element, the row of volumes. This would become the common backbone as it were of general human knowledge. It might take the form of twenty or thirty or forty volumes and it would go to libraries, colleges, schools, institutions, newspaper offices, ministries and so on all over the world. It would be undergoing continual revision. Its various volumes would be in process of replacement, more or less frequently according to the permanence or impermanence of their contents. And from this Standard Encyclopaedia would be drawn a series of text-books and shorter reference encyclopaedias and encyclopaedic dictionaries For individual and casual use.

That crudely is the gist of what I am submitting to you. A double-faced organisation, a perpetual digest and conference on the one hand and a system of publication and distribution on the other. It would be a clearing house for universities and research institutions; it would play the role of a cerebral cortex to these essential ganglia. On the one hand this organisation should be in direct touch with all the original thought and research in the world; on the other it should extend its informing tentacles to every intelligent individual in the community—the new world community. In that little world of the eighteenth century, what we may call the mind of the community scarcely extended below the gentlefolk, the clergy and the professions. There was no primary education for the common man at all. He did not even read. He was a mere toiler. It hardly mattered how little he knew—and the less he thought the better for social order. But machinery abolishes mere toil altogether. The new world has to consist of a world community—say of 2,000 million educated individuals—and the influence of the central encyclopaedic organisation, informing, suggesting, directing, unifying, has to extend to every rank of society and to every corner of the world. The new encyclopaedism is merely the central problem of world education.

Perhaps I should explain that when I speak in this connexion of universities, what I have in mind is primarily assemblies of learned men or men rehearsing their ripe scholarship or conducting original research with such advanced students and student helpers as have been attracted by them and are sharing their fresh and inspiring thoughts and methods. This is a return to the original university idea. The original universities were gatherings of eager people who wanted to know—and who clustered round the teachers who did seem to know. They gathered about these teachers because that was the only way in which they could get their learning. I am talking of that sort of university. That is the primary form of a university. I am not talking here of the collegiate side of a contemporary university, the superficial finishing school exercises of sportive young people mostly of the wealthier classes who don't want to know— young people who mean very little and who have been sent to the university to make useful friendships and get pass-degrees that mean hardly anything at all. These mere finishing-school students are a modern addition, a transitory encumbrance of the halls of learning. I suppose that before very long much of this undergraduate life will merge with the general upward extension of educational facilities to all classes of the community. I assume that the tentacles of this Encyclopaedia we are anticipating, with its comprehensive and orderly supply of knowledge, would intervene beneficially between the specialised research and learning which is the living reality in the university and this really quite modern finishing-school side. The time is rapidly returning when men of outstanding mental quality will consent to teach only such students as show themselves capable of and willing to follow up their distinctive work. The mere graduating crowd with their games and their yells and so forth, will go back to the mere teaching institutions where they properly belong. But I will not spend the few minutes remaining to me upon which is after all a side issue in this discussion. University organisation is not now my subject. I am talking of an essentially new organisation—an addition to the intellectual apparatus of the world. The more important thing now is to emphasise this need—a need the world is likely to realise more and more acutely in the coming years—for such a concentration, which will assemble, co-ordinate and distribute accumulated knowledge. It will link, supplement and no doubt modify profoundly, the universities, schools and other educational organisations we possess already, but it will not in itself be a part of them.

Let me make it perfectly clear that for the present it is desirable to leave this project of a World Encyclopaedic organisation vague—in all but its essential form and function. It might prove disastrous to have it crystallise out prematurely. Such premature crystallisation of a thing needed by the world can produce, we now realise, a rigid obstructive reality, just like enough to our actual requirements to cripple every effort to replace it later by a more efficient organisation. Explicit constitutions for social and political institutions, are always dangerous things if these institutions are to live for any length of time. If a thing is really to live it should grow rather than be made. It should never be something cut and dried. It should be the survivor of a series of trials and fresh beginnings—and it should always be amenable to further amendment.

So that while I believe that ultimately the knowledge systems of the world must be concentrated in this world brain, this permanent central Encyclopaedic organisation with a local habitat and a world-wide range—just as I believe that ultimately the advance of aviation must lead, however painfully and tortuously by way of World Air Control, to the political, economic and financial federation of the world—yet nevertheless I suggest that to begin with, the evocation of this World Encyclopaedia may begin at divergent points and will be all the better for beginning at divergent points.

Of the demand for it, and of the readiness for it in our world today, I have no sort of doubt. Ask the book selling trade. Any books that give or even seem to give, any sort of conspectus of philosophy, of science, of general knowledge, have a sure abundant sale. We have the fullest encouragement for bolder and more strenuous efforts in the same direction. People want this assembling of knowledge and ideas. Our modern community is mind-starved and mind-hungry. It is justifiably uneasy and suspicious of the quality of what it gets. The hungry sheep look up and are not fed—at least they are not fed properly. They want to know, One of the next steps to take, it seems to me, is to concentrate this diffused demand, to set about the definite organisation of a sustained movement, of perhaps a special association or so, to bring a World Encyclopaedia into being. And while on the one hand we have this world-wide receptivity to work upon, on the other hand we have among the men of science in particular a very full realisation of the need for a more effective correlation of their work. It is not only that they cannot communicate their results to the world; they find great difficulty in communicating their results to one another. Among other collateral growths of the League of Nations is a certain Committee of Intellectual Co-operation which has now an official seat in Paris. Its existence shows that even as early as 1919, someone had realised the need for some such synthesis of mental activities as we are now discussing. But in timid, politic and scholarly hands the Committee of Intellectual Co-operation has so far achieved little more than a building, a secretary and a few salaries. The bare idea of a World Encyclopaedia in its present delicate state would give it heart failure. Still there it is, a sort of seed that has still to germinate, waiting for some vitalising influence to stir it to action and growth. And going on at present, among scientific workers, library workers, bibliographers and so forth, there is a very considerable activity for an assembling and indexing of knowledge. An important World Congress of Documentation took place this August in Paris. I was there as an English delegate and I met representatives of forty countries— and my eyes were opened to the very considerable amount of such harvesting and storage that has already been done. From assembling to digesting is only a step—a considerable and difficult step but, none the less, an obvious step. In addition to these indexing activities there has recently been a great deal of experimentation with the microfilm. It seems possible that in the near future, we shall have microscopic libraries of record, in which a photograph of every important book and document in the world will be stowed away and made easily available for the inspection of the student. The British Museum library is making microfilms of the 4,000 books it possesses that were published before 1550 and parallel work is being done here in America. Cheap standardised projectors offer no difficulties. The bearing of this upon the material form of a World Encyclopaedia is obvious. The general public has still to realise how much has been done in this field and how many competent and disinterested men and women are giving themselves to this task. The time is close at hand when any student, in any part of the world, will be able to sit with his projector in his own study at his or her convenience to examine any book, any document, in an exact replica.

Concurrently with this movement towards documentation, we may very possibly have a phase when publishers will be experimenting in the production of larger and better Encyclopaedias, all consciously or unconsciously attempting to realise the final world form. And satisfy a profitable demand. The book salesman from the days of Diderot onward has shown an extraordinary knack for lowering the quality of this sort of enterprise, but I did not see why groups of publishers throughout the world should not presently help very considerably in the beginning of a permanent Encyclopaedic foundation. But such questions of ways and means of distribution belong to a later stage of this great intellectual development which lies ahead of us. I merely glance at them here.

There are certain responses that I have observed crop up almost automatically in people's minds when they are confronted with this project of a worldwide organisation of all that is thought and known. They will say that an Encyclopaedia must always be tendentious and within certain limits—but they are very wide limits— that must be true. A World Encyclopaedia will have by its very nature to be what is called liberal. An Encyclopaedia appealing to all mankind can admit no narrowing dogmas without at the same time admitting corrective criticism. It will have to be guarded editorially and with the utmost jealousy against the incessant invasion of narrowing propaganda. It will have a general flavour of what many people will call scepticism. Myth, however venerated, it must treat as myth and not as a symbolical rendering of some higher truth or any such evasion. Visions and projects and theories it must distinguish from bed-rock fact. It will necessarily press strongly against national delusions of grandeur, and against all sectarian assumptions.

It will necessarily be for and not indifferent to that world community of which it must become at last an essential part. If that is what you call bias, bias the world Encyclopaedia will certainly have. It will have, and it cannot help but have, a bias for organisation, comparison, construction and creation. It is an essentially creative project. It has to be the dominant factor in directing the growth of a new world.

Well, there you have my anticipation of the primary institution which has to appear if that world-wide community towards which mankind, willy-nilly, is being impelled, is ever to be effectively attained. The only alternative I can see is social dissolution and either the evolution of a new, more powerful type of man, or the extinction of our species. I sketch roughly—it seems to be my particular role in life to do these broad sketches and outlines and then stand aside—but I do my best to make the picture plain and understandable. And for me at any rate this is no Utopian dream. It is a forecast, however inaccurate and insufficient, of an absolutely essential part of that world community to which I believe we are driving now. I do not believe there is any emergence for mankind from this age of disorder, distress and fear in which we are living, except by way of such a deliberate vast reorganisation of our intellectual life and our educational methods as I have outlined here. I have been talking of real intellectual forces and foreshadowing the emergence of a vital reality. I have been talking of something which may even be recognizably in active operation within a lifetime—or a lifetime or so, from now—this consciously and deliberately organised brain for all mankind. In a few score years there will be thousands of workers at this business of ordering and digesting knowledge where now you have one. There will be a teacher for every dozen children and schools as unlike the schools of to-day as a liner is unlike the *Mayflower*. There will not be an illiterate left in the world. There will hardly be an uninformed or misinformed person. And the brain of the whole mental network will be the Permanent World Encyclopaedia.

Well, I have designedly put much controversial matter before you, and I have not hesitated to put it in a provocative manner. You will, I know, understand that every new thing is apt to seem crude at first. Forgive my crudities. But my time has been short for what I had to say, and I have said it in the way that seemed most challenging and most likely to produce further discussion.

III. THE IDEA OF A PERMANENT
WORLD ENCYCLOPAEDIA

FIRST PUBLISHED IN HARPER'S MAGAZINE,
APRIL 1937 CONTRIBUTED TO THE NEW
ENCYCLOPÆDIE FRANÇAISE, AUGUST 1937

IT is probable that the idea of an encyclopaedia may undergo very considerable extension and elaboration in the near future. Its full possibilities have still to be realised. The encyclopaedias of the past have sufficed for the needs of a cultivated minority. They were written "for gentlemen by gentlemen" in a world wherein universal education was unthought of, and where the institutions of modern democracy with universal suffrage, so necessary in many respects, so difficult and dangerous in their working, had still to appear. Throughout the nineteenth century encyclopaedias followed the eighteenth-century scale and pattern, in spite both of a gigantic increase in recorded knowledge and of a still more gigantic growth in the numbers of human beings requiring accurate and easily accessible information. At first this disproportion was scarcely noted, and its consequences not at all. But many people now are coming to recognise that our contemporary encyclopaedias are still in the coach-and-horses phase of development, rather than in the phase of the automobile and the aeroplane. Encyclopaedic enterprise has not kept pace with material progress. These observers realise that modern facilities of transport, radio, photographic reproduction and so forth are rendering practicable a much more fully succinct and accessible assembly of fact and ideas than was ever possible before.

Concurrently with these realisations there is a growing discontent with the part played by the universities, schools and libraries in the intellectual life of mankind. Universities multiply, schools of every grade and type increase, but they do not enlarge their scope to anything like the urgent demands of this troubled and dangerous age. They do not perform the task nor exercise the authority that might reasonably be attributed to the thought and knowledge organisation of the world. It is not, as it should be, a case of larger and more powerful universities co-operating more and more intimately, but of many more universities of the old type, mostly ill-endowed and uncertainly endowed, keeping at the old educational level.

Both the assembling and the distribution of knowledge in the world at present are extremely ineffective, and thinkers of the forward-looking type whose ideas we are now considering, are beginning to realise that the most hopeful line for the development of our racial intelligence lies rather in the direction of creating a new world organ for the collection, indexing, summarising and release of knowledge, than in any further tinkering with the highly conservative and resistant university system, local, national and traditional in texture, which already exists. These innovators, who may be dreamers today, but who hope to become very active organisers tomorrow, project a unified, if not a centralised, world organ to "pull the mind of the world together", which will be not so much a rival to the universities, as a supplementary and co-ordinating addition to their educational activities on a planetary scale. The phrase "Permanent World Encyclopaedia" conveys the gist of these ideas. As the core of such an institution would be a world synthesis of bibliography and documentation with the indexed archives of the world. A great number of workers would be engaged perpetually in perfecting this index of human knowledge and keeping it up to date. Concurrently, the resources of micro-photography, as yet only in their infancy, will be creating a concentrated visual record. Few people as yet, outside the world of expert librarians and museum curators and so forth, know how manageable well-ordered facts can be made, however multitudinous, and how swiftly and completely even the rarest visions and the most recondite matters can be recalled, once they have been put in place in a well-ordered scheme of reference and reproduction. The American microfilm experts, even now, are making facsimiles of the rarest books, manuscripts, pictures and specimens, which can then be made easily accessible upon the library screen. By means of the microfilm, the rarest and most intricate documents and articles can be studied now at first hand, simultaneously in a score of projection rooms. There is no practical obstacle whatever now to the creation of an efficient index to all human knowledge, ideas and achievements, to the creation, that is, of a complete planetary memory for all mankind. And not simply an index; the direct reproduction of the thing itself can be summoned to any properly prepared spot. A microfilm, coloured where necessary, occupying an inch or so of space and weighing little more than a letter, can be duplicated from the records and sent anywhere, and thrown enlarged upon the screen so that the student may study it in every detail.

This in itself is a fact of tremendous significance. It

foreshadows a real intellectual unification of our race. The whole human memory can be, and probably in a short time will be, made accessible to every individual. And what is also of very great importance in this uncertain world where destruction becomes continually more frequent and unpredictable, is this, that photography affords now every facility for multiplying duplicates of this —which we may call?—this new all-human cerebrum. It need not be concentrated in any one single place. It need not be vulnerable as a human head or a human heart is vulnerable. It can be reproduced exactly and fully, in Peru, China, Iceland, Central Africa, or wherever else seems to afford an insurance against danger and interruption. It can have at once, the concentration of a craniate animal and the diffused vitality of an amoeba.

This is no remote dream, no fantasy. It is a plain statement of a contemporary state of affairs. It is on the level of practicable fact. It is a matter of such manifest importance and desirability for science, for the practical needs of mankind, for general education and the like, that it is difficult not to believe that in quite the near future, this Permanent World Encyclopaedia, so compact in its material form and so gigantic in its scope and possible influence, will not come into existence.

Its uses will be multiple and many of them will be fairly obvious. Special sections of it, historical, technical, scientific, artistic, e.g. will easily be reproduced for specific professional use. Based upon it, a series of summaries of greater or less fullness and simplicity, for the homes and studies of ordinary people, for the college and the school, can be continually issued and revised. In the hands of competent editors, educational directors and teachers, these condensations and abstracts incorporated in the world educational system, will supply the humanity of the days before us, with a common understanding and the conception of a common purpose and of a commonweal such as now we hardly dare dream of. And its creation is a way to world peace that can be followed without any very grave risk of collision with the warring political forces and the vested institutional interests of today. Quietly and sanely this new encyclopaedia will, not so much overcome these archaic discords, as deprive them, steadily but imperceptibly, of their present reality. A common ideology based on this Permanent World Encyclopaedia is a possible means, to some it seems the only means, of dissolving human conflict into unity.

This concisely is the sober, practical but essentially colossal objective of those who are seeking to synthesise human mentality today, through this natural and reasonable development of encyclopaedism into a Permanent World Encyclopaedia.

IV. PASSAGE FROM A SPEECH
TO THE CONGRES MONDIAL

DE LA DOCUMENTATION
UNIVERSELLE, PARIS AUGUST 20TH, 1937

IT is dawning upon us, we lay observers, that this work of documentation and bibliography, is in fact nothing less than the beginning of a world brain, a common world brain. What you are making me realise is a sort ofcerebrum for humanity, a cerebral cortex which (when it is fully developed) will constitute a memory and a perception of current reality For the entire human race. Plainly we have to make it a centralised and uniform organisation but, as Mr. Watson Davis is here to remind us, it need not have any single local habitation because the continually increasing facilities of photography render reduplication of our indices and records continually easier. In these days of destruction, violence and general insecurity, it is comforting to think that the brain of mankind, the race brain, can exist in numerous replicas throughout the world.

At first our activities are necessarily receptive and we begin most easily with the documentation of concrete facts, and I do not see how this new and great encyclopaedia, this race brain, can develop into anything but a great structure for the comparison, reconciliation and synthesis of common guiding ideas for the whole world. What is gathered will be digested and the results returned through the channels of education, literature and the press to the whole world.

Please do not imagine that I am indulging in any fantasy when I talk of your work and your accumulations as the rudimentary framework of a world brain. I am speaking of a process of mental organisation throughout the world which I believe to be as inevitable as anything can be in human affairs. The world has to pull its mind together, and this is the beginning of its effort. The world is a Phoenix. It perishes in flames and even as it dies it is born again. This synthesis of knowledge is the necessary beginning to the new world.

It is good to be meeting here in Paris where the first encyclopaedia of ideas was made. It is good for representatives from forty countries to be breathing the clear, comprehensive and systematic atmosphere of France, to be recreating themselves in the presence of its sympathetic creative understanding. Again I would thank our hosts for bringing this congress together and enabling a number of widely scattered workers to realise something of the real greatness of the task to which they have devoted themselves.

V. THE INFORMATIVE
CONTENT OF EDUCATION

THE PRESIDENTIAL ADDRESS TO THE EDUCATIONAL
SCIENCE SECTION OF THE BRITISH ASSOCIATION
FOR THE ADVANCEMENT OF SCIENCE,
GIVEN ON SEPTEMBER 2ND, 1937,

AT NOTTINGHAM, AS READ BY MR, WELLS

SECTION L of the British Association is of necessity one of the least specialised of all sections. Its interests spread far beyond professional limitations. It is a section where anyone who is so to speak a citizen at [age may hope to play a part that is not altogether an impertinent intrusion. And it is in the character of a citizen at large that I have accepted the very great honour that you have offered me in making me the President of this Section. I have no other claim whatever upon your attention. Since the remote days when as a needy adventurer I taught as non-resident master in a private school, invigilated at London University examinations, raided the diploma examinations of the College of Preceptors for the money prises offered, and, in the most commercial spirit, crammed candidates for the science examinations of the University, I have spent very few hours indeed in educational institutions. Most of those were spent in the capacity of an inquiring and keenly interested parent at Oundle School. I doubt if there is any member of this section who has not had five times as much teaching experience as I have, and who is not competent to instruct me upon all questions of method and educational organisation and machinery. So I will run no risks by embarking upon questions of that sort. But on the other hand, if I know very little of educational methods and machinery I have had a certain amount of special experience in what those methods produce and what the machinery turns out.

I have been keenly interested for a number of years, and particularly since the War, in public thought and public reactions, in what people know and think and what they are ready to believe. What they know and think and what they are ready to believe impresses me as remarkably poor stuff. A general ignorance—even in respectable quarters—of some of the most elementary realities of the political and social life of the world is, I believe, mainly accountable for much of the discomfort and menace of our times. The uninstructed public intelligence of our community is feeble and convulsive. It is still a herd intelligence. It tyrannises here and yields to tyranny there. What is called elementary education throughout the world does not in fact educate, because it does not properly inform. I realised this very acutely during the later stages of the War and it has been plain in my mind ever since. It led to my taking an active part in the production of various outlines and summaries of contemporary knowledge. Necessarily they had the defects and limitations of a private adventure, but in making them I learnt a great deal about—what shall I say—the contents of the minds our schools are turning out as taught.

And so now I propose to concentrate the attention of this

Section for this meeting on the question of what is taught as fact, that is to say upon the informative side of educational work. For this year I suggest we give the questions of drill, skills, art, music, the teaching of languages, mathematics and other symbols, physical, aesthetic, moral and religious training and development, a rest, and that we concentrate on the inquiry: What are we telling young people directly about the world in which they are to live? What is the world picture we are presenting to their minds? What is the framework of conceptions about reality and about obligation into which the rest of their mental existences will have to be fitted? I am proposing in fact a review of the informative side of education, wholly and solely-informative in relation to the needs of modern life.

And here the fact that I am an educational outsider—which in every other relation would be a disqualification—gives me certain very real advantages. I can talk with exceptional frankness. And I am inclined to think that in this matter of the informative side of education frankness has not always been conspicuous. For what I say I am responsible only to the hearer and my own self-respect. I occupy no position from which I can be dismissed as unsound in my ideas. I follow no career that can be affected by anything I say. I follow, indeed, no career. That's all over. I have no party, no colleagues or associates who can be embarrassed by any unorthodox suggestions I make. Every schoolmaster, every teacher, nearly every professor must, by the nature of his calling, be wary, diplomatic, compromising; he has his governors to consider, his college to consider, his parents to consider, the local press to consider; he must not say too much nor say anything that might be misinterpreted and misunderstood. I can. And so I think I can best serve the purposes of the British Association and this section by taking every advantage of my irresponsibility, being as unorthodox and provocative as I can be, and so possibly saying a thing or two which you are not free to say but which some of you at any rate will be more or less willing to have said.

Now when I set myself to review the field of inquiry I have thus defined, I found it was necessary to take a number of very practical preliminary issues into account. As educators we are going to ask what is the subject-matter of a general education? What do we want known? And how do we want it known? What is the essential framework of knowledge that should be established in the normal citizen of our modern community? What is the irreducible minimum of knowledge for a responsible human being today? I say irreducible minimum—and I do so, because I know at least enough of school work to know the grim significance of the school time-table and of the leaving school age. Under contemporary conditions our only prospect of securing a mental accord throughout the community is by laying a common foundation of knowledge and ideas in the school years. No one believes today, as our grandparents—perhaps for most of you it would be better to say great-grandparents—believed, that education had an end somewhen about adolescence. Young people then leave school or college under the imputation that no one could teach them any more. There has been a quiet but complete revolution in people's ideas in this respect and now it is recognised almost universally that people in a modern community must be learners to the end of their days. We shall be giving a considerable amount of attention to continuation, adult and post-graduate studies in this section, this year. It would be wasting our opportunities not to do so. Here in Nottingham University College we have the only Professorship in Adult

Education in England, and under Professor Peers the Adult Education Department which is in close touch with the Workers' Educational Association has broadened its scope far beyond the normal range of Adult Education. Our modern ideas seem to be a continuation of learning not only for university graduates and practitioners in the so-called intellectual professions, but for the miner, the plough-boy, the taxi-cab driver, and the out-of-work, throughout life. Our ultimate aim is an entirely educated population.

Nevertheless it is true that what I may call the main beams and girders of the mental framework must be laid down, soundly or unsoundly, before the close of adolescence. We live under conditions where it seems we are still only able to afford for the majority of our young people, freedom from economic exploitation, teachers even of the cheapest sort and some educational equipment, up to the age of 14 or 15, and we have to fit our projects to that. And even if we were free to carry on with unlimited time and unrestrained teaching resources, it would still be in those opening years that the framework of the mind would have to be made.

We have got to see therefore that whatever we propose as this irreducible minimum of knowledge must be imparted between infancy and—at most, the fifteenth or sixteenth year. Roughly, we have to get it into ten years at the outside.

And next let us turn to another relentlessly inelastic-seeming case—and that is, the school time-table. How many hours in the week have we got for this job in hand? The maximum school hours we have available are something round about thirty, but out of this we have to take time for what I may call the non-informative teaching, teaching to read, teaching to write clearly, the native and foreign language teaching, basic mathematical work, drawing, various forms of manual training, music and so forth. A certain amount of information may be mixed in with these subjects but not very much. They are not what I mean by informative subjects. By the time we are through with these non-informative subjects, I doubt if at the most generous estimate we can apportion more than six hours a week to essentially informative work. Then let us, still erring on the side of generosity, assume that there are forty weeks of schooling in the year. That gives us a maximum of 240 hours in the year. And if we take ten years of schooling as an average human being's preparation for life, and if we disregard the ravages made upon our school time by measles, chicken-pox, whooping-cough, coronations and occasions of public rejoicing, we are given 2,400 hours as all that we can hope for as our time allowance for building up a coherent picture of the world, the essential foundation of knowledge and ideas, in the minds of our people. The complete framework of knowledge has to be established in 200 dozen hours. It is plain that a considerable austerity is indicated for us. We have no time to waste, if our schools are not to go on delivering, year by year, fresh hordes of fundamentally ignorant, unbalanced, uncritical minds, at once suspicious and credulous, weakly gregarious, easily baffled and easily misled, into the monstrous responsibilities and dangers of this present world. Mere cannon-fodder and stuff for massacres and stampedes.

Out question becomes therefore: "What should people know—whatever else they don't know? Whatever else we may leave over—for leisure-time reading, for being picked up or studied afterwards—what is the irreducible minimum that we ought to teach as clearly, strongly and conclusively as we know how?"

And now I—and you will remember my role is that of the irresponsible outsider, the citizen at large—I am going to set before you one scheme of instruction for your consideration. For it I demand all those precious 2,400 hours. You will perceive, as I go on, the scheme is explicitly exclusive of several contradictory and discursive subjects that now find a place in most curricula, and you will also find doubts arising in your mind about the supply and competence of teachers, a difficulty about which I hope to say something before my time is up. But teachers are for the world and not the world for teachers. If the teachers we have today are not equal to the task required of them, then we have to recondition our teachers or replace them. We live in an exacting world and a certain minimum of performance is required of us all. If children are not to be given at least this minimum of information about the world into which they have come—through no fault of their own—then I do think it would be better for them and the world if they were not born at all. And to make what I have to say as clear as possible I have had a diagram designed which I will unfold to you as my explanation unfolds. You have already noted I have exposed the opening stage of my diagram. You see I make a three-fold division of the child's impressions and the matters upon which its questions are most lively and natural. I say nothing about the child learning to count, scribble, handle things, talk and learn the alphabet and so forth because all these things are ruled out by my restriction of my address to information only. Never mind now what it wants to do—or wants to feel. This is what it wants to know. In all these educational matters, there is, of course, an element of overlap. As it learns about things and their relationship and interaction its vocabulary increases and its ideas of expression develop. You will make an allowance for that.

[chart]

And now I bring down my diagram to expose the first stage of positive and deliberate teaching. We begin telling true stories of the past and of other lands. We open out the child's mind to a realisation that the sort of life it is living is not the only life that has been lived and that human life in the past has been different from what it is today and on the whole that it has been progressive. We shall have to teach a little about law and robbers, kings and conquests, but I see no need at this stage to afflict the growing mind with dates and dynastic particulars. I hope the time is not Far distant when children even of 8 or 9 will be freed altogether from the persuasion that history is a magic recital beginning "William the Conqueror, 1066". Much has been done in that direction. Much remains to be done. Concurrently, we ought to make the weather and the mud pie our introduction to what Huxley christened long ago Elementary Physiography. We ought to build up simple and clear ideas from natural experience.

We start a study of the states of matter with the boiling, evaporation, freezing and so on of water and go on to elementary physics and chemistry. Local topography can form the basis of geography. We shall have to let our learner into the secret that the world is a globe-and for a time I think that has to be a bit of dogmatic teaching. It is not so easy as many people suppose to prove that the world is spherical and that proof may very well be left to make an exercise in logic later on in education. Then comes biology. Education I rejoice to see is rapidly becoming more natural, more biological. Most young children are ready to learn a great deal more than most teachers can give them about animals. I think we might easily turn the bear, the wolf] the tiger and the ape from holy terrors and nightmare material into sympathetic creatures, if we brought some realisation of how these creatures live, what their real excitements are, how they are sometimes timid, into the teaching. I don't think that descriptive botany is very suitable for young children. Flowers and leaves and berries are bright and attractive, a factor in aesthetic education, but I doubt if, in itself, vegetation can hold the attention of the young. Sometimes I think we bore young children with premature gardens. But directly we begin to deal with plants as hiding-places, homes and food for birds and beasts, the little boy or girl lights up and learns. And with this natural elementary zoology and botany we should begin elementary physiology. How plants and animals live, and what health means for them. There I think you have stuff enough for all the three or four hundred hours we can afford for the foundation stage of knowledge. Outside this substantial teaching of school hours the child will be reading and indulging in imaginative play—and making that clear distinction children do learn to make between truth and fantasy—about fairyland, magic carpets and seven-league boots, and all the rest of it. So far as my convictions go I think that the less young children have either in or out of school I of what has hitherto figured as history, the better. I do not see either the charm or the educational benefit of making an important subject of and throwing a sort of halo of prestige and glory about the criminal history of royalty, the murder of the Princes in the Tower, the wives of Henry VIII, the families of Edward I and James I, the mistresses of Charles II, Sweet Nell of Old Drury, and all the rest of it. I suggest that the sooner we get all that unpleasant stuff out of schools, and the sooner that we forget the border bickerings of England, France,

Scotland, Ireland and Wales, Bannockburn, Flodden, Crécy and Agincourt, the nearer our world will be to a sane outlook upon life. In this survey of what a common citizen should know I am doing my best to elbow the scandals and revenges which once passed as English history into an obscure corner or out of the picture altogether. But I am not proposing to eliminate history from education—far from it. Let me bring down my diagram a stage further and you will see how large a proportion of our treasure of 2,400 hours I am proposing to give to history. This next section represents about 800 to 1,000 pre-adolescent hours. It is the schoolboy-schoolgirl stage. And here the history is planned to bring home to the new generation the reality that the world is now one community. I believe that the crazy combative patriotism that plainly threatens to destroy civilisation today is very largely begotten in their school history lessons. Our schools take the growing mind at a naturally barbaric phase and they inflame and fix its barbarism. I think we underrate the formative effect of this perpetual reiteration of how we won, how our Empire grew and how relatively splendid we have been in every department of life. We are blinded by habit and custom to the way it infects these growing minds with the chronic and nearly incurable disease of national egotism. Equally mischievous is the furtive anti-patriotism of the leftish teacher. I suggest that we take on our history from the simple descriptive anthropology of the elementary stage to the story of the early civilisations.

We are dealing here with material that was not even available for the schoolmasters and mistresses who taught our fathers. It did not exist. But now we have the most lovely stuff to hand, far more exciting and far more valuable than the quarrels of Henry II and Becket or the peculiar unpleasantnesses of King James or

King John. Archaeologists have been piecing together a record of the growth of the primary civilisations and the developing roles of priest, king, farmer, warrior, the succession of stone and copper and iron, the appearance of horse and road and shipping in the expansions of those primordial communities. It is a far finer story to tell a boy or girl and there is no reason why it should not be told. Swinging down upon these early civilisations came first the Semitic-speaking peoples and then the Aryan-speakers, Persian, Macedonian, Roman, followed one another, Christendom inherited from Rome and Islam from Persia, and the world began to assume the shapes we know today. This is great history and also in its broad lines it is a simple history—upon it we can base a lively modern intelligence, and now it can be Put in a form just as comprehensible and exciting for the school phase as the story of our English kings and their territorial, dynastic and sexual entanglements. When at last we focus our attention on the British Isles and France we shall have the affairs of these regions in a proper proportion to the rest of human adventure. And our young people will be thinking less like gossiping court pages and more like horse-riders, seamen, artist-artisans, roadmakers, and city-builders, which I take it is what in spirit we want them to be. Measured by the great current of historical events, English history up to quite recent years is mere hole-and-corner history.

And I have to suggest another exclusion. We are telling our young people about the real past, the majestic expansion of terrestrial events. In these events the little region of Palestine is no more than a part of the highway between Egypt and Mesopotamia. Is there any real reason nowadays for exaggerating its importance in the past? Nothing really began there, nothing was worked out there. All the historical part of the Bible abounds in wild exaggeration of the importance of this little strip of land. We were all brought up to believe in the magnificence of Solomon's temple and it is a startling thing for most of us to read the account of its decorations over again and tum its cubits into feet. It was smaller than most barns. We all know the peculiar delight of devout people when, amidst the endless remains of the great empires of the past, some dubious fragment is found to confirm the existence of the Hebrews. Is it not time that we recognised the relative historical insignificance of the events recorded in Kings and Chronicles, and ceased to throw the historical imagination of our young people out of perspective by an over-emphasised magnification of the national history of Judea; To me this lack of proportion in our contemporary historical teaching seems largely responsible for the present troubles of the world. The political imagination of our times is a hunch-back imagination bent down under an exaggeration. It is becoming a matter of life and death to the world to straighten that backbone and reduce that frightful nationalist hunch.

Look at our time-table and what we have to teach. If we give history four-tenths of all the time we have for imparting knowledge at this stage that still gives us at most something a little short of 400 hours altogether. Even if we think it desirable to perplex another generation with the myths of the Creation, the Flood, the Chosen People and so forth, even if we want to bias it politically with tales of battles and triumphs and ancient grievances, we haven't got the time for it—any more than we have the time for the really quite unedifying records of all the Kings and Queens of England and their claims on this and that. So far as the school time-table goes we are faced with a plain alternative. One thing or the other. Great history and hole-in-corner history?

The story of mankind or the narrow, self-righteous, blinkered stories of the British Islands and the Jews? There is a lot more we have to put into the heads of our young people over and above History. It is the main subject of instruction but even so, it is not even half of the informative work that ought to be got through in this school stage. We have to consider the collateral subject of geography and a general survey of the world. We want to see our world in space as well as our world in time. We may have a little map-making here, but I take it what is needed most are reasonably precise ideas of the various types of country and the distinctive floras and faunas of the main regions of the world. We do not want our budding citizens to chant lists of capes and rivers, but we do want them to have a real picture in their minds of the Amazon forests, the pampas, the various phases in the course of the Nile, the landscape of Labrador mountains, and so on, and also we want something like a realisation of the sort of human life that is led in these regions. We have enormous resources now in cheap photography, in films, and so forth, that even our fathers never dreamt of—to make all this vivid and real. New methods are needed to handle these new instruments, but they need not be overwhelmingly costly. And also our new citizen should know enough of topography to realise why London and Rio and New York and Rome and Suez happen to be where they are and what sort of places they are.

Geography and History run into each other in this respect and, on the other hand, Geography reaches over to Biology. Here again our schools lag some fifty years behind contemporary knowledge. The past half-century has written a fascinating history of the succession of living things in time and made plain all sorts of processes in the prosperity, decline, extinction, and replacement of species. We can sketch the wonderful and inspiring story of life now from its beginning. Moreover, we have a continually more definite account of the sequence of sub-man in the world and the gradual emergence of our kind. This is elementary, essential, interesting and stimulating stuff for the young, and it is impossible to consider anyone a satisfactory citizen who is still ignorant of that great story.

And finally, we have the science of inanimate matter. In a world of machinery, optical instruments, electricity, radio and so forth we want to lay a sound foundation of pure physics and chemistry upon the most modern lines—for everyone. Some of this work will no doubt overlap the mathematical teaching and the manual training and steal a little badly needed time from them. And finally, to meet awakening curiosity and take the morbidity out of it, we shall have to tell our young people and especially our young townspeople, about the working of their bodies, about reproduction and about the chief diseases, enfeeblements and accidents that lie in wait for them in the world.

That, I think, completes my summary of all the information we can hope to give in the lower school stage. And as I make it I am acutely aware of your unspoken comment. With such teachers as we have—teachers trained only to reaction, overworked, underpaid, hampered by uninspiring examinations, without initiative, without proper leisure. Young and inexperienced or old and discouraged. You may do this sort of thing, here and there, under favourable conditions, with the splendid élite of the profession, the 10 per cent who are interested, but not as a general state of affairs. Well, I think that it is a better rule of life, first to make sure of what you want and then set about getting it, rather than to consider what you can easily, safely and meanly get, and then set about reconciling yourself to it. I admit we cannot have a modern education without a modernised type of teacher. A teacher enlarged and released. Many of our teachers—and I am not speaking only of elementary schools—are shockingly illiterate and ignorant. Often they know nothing but school subjects; sometimes they scarcely know them. Even the medical profession does not present such extremes between the discouraged routine worker and the enthusiast. Everything I am saying now implies a demand for more and better teachers-better paid, with better equipment. And these teachers will have to be kept fresh. It is stipulated in most leases that we should paint our houses outside every three years and inside every seven years, but nobody ever thinks of doing up a school teacher. There are teachers at work in this country who haven't been painted inside for fifty years. They must be damp and rotten and very unhealthy for all who come in contact with them. Two-thirds of the teaching profession now is in urgent need of being either reconditioned or superannuated. In this advancing world the reconditioning of both the medical and the scholastic practitioner is becoming a very urgent problem indeed, but it is not one that I can deal with here. Presently this section will be devoting its attention to adult education and then I hope the whole question of professional and technical refreshment will be ventilated.

And there is another matter also closely allied to this question of the rejuvenation of teachers, at which I can only glance now, and that is the bringing of school books up to date. In this informative section of school work there is hardly a subject in which knowledge is not being vigorously revised and added to. But our school work does not follow up the contemporary digesting of knowledge. Still less do our school libraries. They are ten, fifteen years out of date with much of their information. Our prison libraries by the by are even worse. I was told the other day of a virtuous prisoner who wanted to improve his mind about radio. The prison had a collection of technical works made for such an occasion and the latest book on radio was dated 1920. There is, I have been told, an energetic New School Books Association at work in this field, doing what it can to act in concert with those all too potent authorities who frame our examination syllabuses. I am all for burning old school books. Some day perhaps we shall have school books so made that at the end of ten or twelve years, let us say, they will burst into flames and inflict severe burns upon any hands in which they find themselves. But at present that is a little Utopian. It is even more applicable to the next stage of knowledge to which we are now coming.

This stage represents our last 1,000 hours and roughly I will call it the upper form or upper standard stage. It is really the closing phase of the available school period. Some of the matter I have marked for the history of this grade might perhaps be given i.n Grade B and vice versa. We have still a lot to do if we are to provide even a skeleton platform for the mind of our future citizen. He has still much history to learn before his knowledge can make an effective contact with his duties as a voter.

You see I am still reserving four-tenths of the available time, that is to say nearly 400 hours for history. But now we are presenting a more detailed study of such phenomena as the rise and fall of the Ottoman Empire, the rise of Russia, the history of the Baltic, the rise and fall of the Spanish power, the Dutch, the first and second British Empires, the belated unifications of Germany and Italy. Then as I have written we want our modern citizen to have some grasp of the increasing importance of economic changes in history and the search for competent economic direction and also of the leading theories of individualism, socialism, the corporate state, communism.

For the next five-and-twenty years now the ordinary man all over the earth will be continually confronted with these systems of ideas. They are complicated systems with many implications and applications. Indeed they are aspects of life rather than systems of ideas. But we send out our young people absolutely unprepared for the heated and biased interpretations they will encounter. We hush it up until they are in the thick of it. And can we complain of the consequences? The most the poor silly young things seem able to make of it is to be violently and self-righteously Anti- something or other, Anti-Red, Anti-Capitalist, Anti-Fascist. The more ignorant you are the easier it is to be an Anti. To hate something without having something substantial to put against it. Blame something else. A special sub-section of history in this grade should be a course in the history of war, which is always written and talked about by the unwary as though it had always been the same thing, while as a matter of fact—except for its violence—it has changed profoundly with every change in social, political and economic life. Clearly parallel to this history our young people need now a more detailed and explicit acquaintance with world geography, with the different types of population in the world and the developed and undeveloped resources of the globe. The devastation of the world's forests, the replacement of pasture by sand deserts through haphazard cultivation, the waste and exhaustion of natural resources, coal, petrol, water, that is now going on, the massacre of important animals, whales, penguins, seals, food fish, should be matters of universal knowledge and concern.

Then our new citizens have to understand something of the broad elements in our modern social structure. They should be given an account of the present phase of communication and trade, of production and invention and above a.l.l they need whatever plain knowledge is available about the conventions of property and money. Upon these interrelated conventions human society rests, and the efficiency of their working is entirely dependent upon the general state of mind throughout the world. We know now that what used to be called the inexorable laws of political economy and the laws of monetary science, are really no more than rash generalisations about human behaviour, supported by a maximum of pompous verbiage and a minimum of scientific observation. Most of our young people come on to adult life, to employment, business and the rest of it, blankly ignorant even of the way in which money has changed slavery and serfdom into wages employment and of how its fluctuations in value make the industrial windmills spin or flag. They are not even warned of the significance of such words as inflation or deflation, and so the wage-earners are the helpless prey at every turn towards prosperity of the savings-snatching financier. Any plausible monetary charlatan can secure their ignorant votes. They

Informative Content of Education 8 5 know no better. They cannot help themselves. Yet the subject of property and money— together they make one subject because money is only the fluid form of property—is scarcely touched upon in any stage in the education of any class in our community. They know nothing about it; they are as innocent as young lambs and born like them for shearing.

And now here you will see I have a very special panel. This I have called Personal Sociology. Our growing citizen has reached an age of self- consciousness and self-determination. He is on the verge of adolescence. He has to be initiated. Moral training does not fall within the scope of the informative content of teaching. Already the primary habits of truthfulness, frankness, general honesty, communal feeling, helpfulness and generosity will or will not have been fostered and established in the youngster's mind by the example of those about him. A mean atmosphere makes mean people, a too competitive atmosphere makes greedy, self-glorifying people, a cruel atmosphere makes fierce people, but this issue of moral tone does not concern us now here. But it does concern us that by adolescence the time has arrived for general ideas about one's personal relationship to the universe to be faced. The primary propositions of the chief religious and philosophical interpretations of the world should be put as plainly and impartially as possible before our young people. They will be asking those perennial questions of adolescence— whence and why and whither. They will have to face, almost at once, the heated and exciting propagandas of theological and sceptical partisans—pro's and anti's. So far as possible we ought to provide a ring of cleat knowledge for these inevitable fights. And also, as the more practical aspect of the question, What am I to do with my life? I think we ought to link with our general study of social structure a study of social types which will direct attention to the choice of a métier. In what spirit will you face the world and what sort of a job do you feel like? This subject of Personal Sociology as it is projected here is the informative equivalent of a confirmation class. It says to everyone: "There are the conditions under which you face your world." The response to these questions, the determination of the will, is however not within our present scope. That is a matter for the religious teacher, for intimate friends and for the inner impulses of the individual. But our children must have the facts.

Finally, you will see that I have apportioned some time, roughly two-tenths of our 1,000 hours, in this grade to the acquisition of specialised knowledge. Individuality is becoming conscious of itself and specialisation is beginning.

Thus I budget, so to speak, for our 2,400 hours of informative teaching. We have brought our young people to the upper form, the upper standard. Most of them arc now going into employment or special training and so taking on a role in the collective life. But there remain some very essential things which cannot be brought into school teaching, not through any want of time, but because of the immaturity of the growing mind. It we are to build a real modern civilisation we must go on with definite informative instruction into and even beyond adolescence. Children and young people are likely to be less numerous proportionally in the years ahead of us in all the more civilised populations and we cannot afford to consume them in premature employment after the fashion of the preceding centuries. The average age of our population is rising and this involves an upward extension of education. And so you will see I suggest what I call an undergraduate or continuation school, Grade D, the upper adolescent stage, which I presume will extend at last to every class in the population, in which at least half the knowledge acquired will be specialised in relation to interest, aptitude and the social needs of the individual. But the other half will still have to be unspecialised, it will have to be general political education. Here particularly comes in that education for citizenship to which this Educational Section is to give attention later. It seems to me altogether preposterous that nowadays our educational organisation should turn out new citizens who are blankly ignorant of the history of the world during the last twenty-five years, who know nothing of the causes and phases of the Great War and are left to the tender mercies of freakish newspaper proprietors and party organisers for their ideas about the world outlook, upon which their collective wills and actions must play a decisive part.

Social organisation is equally a matter for definite information. "We are all socialists nowadays." Everybody has been repeating that after the late Lord Rosebery for years and years. Each for all and all for each. We are all agreed upon the desirability of the spirit of Christianity and of the spirit of Democracy, and that the general interest of the community should not be sacrificed to Private Profit. Yes—beautiful, but what is not realised is that Socialism in itself is little more than a generalisation about the undesirability of irresponsible ownership and that the major problem before the world is to devise some form of administrative organisation that will work better than the scramble of irresponsible owners. That form of administrative organisation has not yet been devised. You cannot expropriate the private adventurer until you have devised a competent receiver for the expropriated industry or service. This complex problem of the competent receiver is the underlying problem of most of our constructive politics. It is imperative that every voter should have some conception of the experiments in economic control that are in progress in Great Britain, the United States of America, Italy, Germany, Russia, and elsewhere. Such experiments are going to affect the whole of his or her life profoundly. So, too, are the experiments in monetary and financial organisation. Many of the issues involved go further than general principles. They are quantitative issues, questions of balance and more or less. A certain elementary training in statistical method is becoming as necessary for anyone living in this world of today as reading and writing. I am asking for this much contemporary history as the crowning phase, the graduation phase of our knowledge-giving. After that much foundation, the informative side of education may well be left to look after itself.

Speaking as a teacher of sorts myself, to a gathering in which teachers probably predominate, I need scarcely dilate upon the fascination of diagram drawing. You will understand how reluctant I was to finish of at Grade D and how natural it was to extend my diagram to two more grades and make it a diagram of the whole knowledge organisation of a modern community. Here then is Grade E, the adult learning that goes on now right through life, keeping oneself up to date, keeping in touch with the living movements about us. I have given a special line to those reconditioning courses that must somehow be made a normal part in the lives of working professional men. It is astonishing how stale most middle-aged medical men, teachers and solicitors

And beyond Grade E I have put a further ultimate grade for the fully adult human being. He or she is learning now, no longer only from books and newspapers and teachers, though there has still to be a lot of that, but as a worker with initiative, making experiments, learning from new experience, an industrialist, an artist, an original writer, a responsible lawyer, an administrator, a statesman, an explorer, a scientific investigator. Grade F accumulates, rectifies, changes human experience. And here I bring in an obsession of mine with which I have dealt before the Royal Institution and elsewhere. You see, indicated by these arrows, the rich results of the work of Grade F flowing into a central world-encyclopaedic-organisation, where it will be continually summarised, clarified, and whence it will be distributed through the general information channels of the world.

So I complete my general scheme of the knowledge organisation of a modern community and submit it to you for your consideration.

I put it before you in good faith as a statement of my convictions. I do not know how it will impress you and I will not anticipate your criticisms. It may seem impossibly bold and "Utopian." But we are living in a world in which a battleship costs £8,000,000, in which we can raise an extra £400,000,000 for armaments with only a slight Stock Exchange qualm, and which has seen the Zeppelin, the radio, the bombing aeroplane come absolutely out of nothing since 1900. And our schools are going along very much as they were going along thirty-seven years ago.

There is only one thing I would like to say in conclusion. Please do me the justice to remember that this is a project for Knowledge Organisation only and solely. It is not an entire scheme of education I am putting before you. It is only a part and a limited part of education—the factual side of education—I have discussed. There are 168 hours in a week and I am dealing with the use of rather less than six during the school year of less than forty weeks—for ten years. It is no good saying as though it was an objection either to my paper or to me, that I neglect or repudiate spiritual, emotional and aesthetic values. They are not disregarded, but they have no place at all in this particular part of the educational scheme. I have said nothing about music, dancing, drawing, painting, exercise and so on and so forth. Not because I would exclude them from education but because they do not fall into the limits of my subject. You no more want these lovely and elementary things mixed up with a conspectus of knowledge than you want playfulness in an ordnance map or perplexing whimsicality on a clock face. You have the remaining 162 hours a week for all that. But the spiritual, emotional, aesthetic lives our children are likely to lead, will hardly be worth living unless they are sustained by such a clear, full and sufficient backbone of knowledge as I have ventured to put before you here.

APPENDIX I. RUFFLED TEACHERS

PUBLISHED IN THE SUNDAY CHRONICLE, SEPTEMBER 12TH, 1937

THE BREEZE AT THE BRITISH ASSOCIATION

I FIND myself in trouble with a number of indignant school inspectors and teachers throughout the country, and I am in the awkward position of a man who is essentially right but who has nevertheless been rather tactless in his phrasing. And also I have suffered from the necessity reporters and editors are under to compress and point what one has said. At times their sense of drama leads them to omit the meat and give only the salt and mustard, and so one's remarks are presented in a state of exaggerated pungency. I find I have wounded more than I intended.

Moreover the reporters at the meeting were supplied with a first draft of my address and this I had already modi6ed in certain respects before I read it. For example, I did not say our schools are "drooling along" much as they did in 1900. "Drooling" was a hard offensive word. Actually I said they were "going along" much as they did in 1900, but the reporters sitting under me with the printed draft before them did not note the change nor did they notice the insertion of a considerable passage upon the underpayment, overwork, limitation upon initiative and so forth that prevent a vigorous teacher from doing himself or herself full justice. But these are subsidiary matters.

I still maintain when all allowances have been made for the progress of education in the past third of a century that elementary teaching throughout the world—even in our own urban elementary schools where progress has been most marked—has not kept pace with the demands of a time in which such things as aviation and radio have leapt out of nothingness into primary importance in our affairs, and in which human power for good or evil has been stupendously increased. Schools as a whole may be going forwards, but nevertheless they are being outrun. In the race between education and catastrophe, catastrophe is winning.

And anxious though I am to salve the feelings of teachers and scholastic authorities in this matter, I am obliged to remark upon one or two characteristics of this storm of protest and repudiation I have provoked. The first is the solidarity of the teachers in their indignation. I say there are teachers who are not up to their job, that some of them have not been done up inside for fifty years. They are as damp and rotten as old houses. And surely every teacher knows that that is true. "Some" is not "all." But will they admit it? Instead they flare up. "You say we are all damp and rotten!" I don't. And when I say two-thirds of the teaching profession is in urgent need of either reconditioning or superannuation, I mean two-thirds and not the whole.

In spite of the anger I have evoked I stick to that rough estimate. The level of qualification is still far too low for modern needs, the amount of reconditioning in brief "holiday" courses and so forth is not enough, and the educational engine in the social apparatus is not up to the stresses it has to meet, We want more and better teachers. We want them urgently. Elementary education lags—throughout the world. I stick to that.

From all the Parts of the country come "retorts" to my address. "Smithstown Director of Education Hits Out", and that sort of thing. Maybe sometimes the journalist barbs the shaft. The more substantial counter-attack is that I am out of date—usually it is "sadly out of date". Poor Mr. Wells! A charming head girl (with photograph) is produced to say that surely I must be thinking of my own school days sixty years ago, and inspectors, masters, head-mistresses and assistants combine in being "amused" at my unawareness of the blinding light in which their pupils live nowadays. They say I have not been in a school for fifty years, which is not exactly true and that lays them open to the obvious remark that some of them seem never to have been out of school all their lives. It is a peculiar atmosphere for I the teacher, that school atmosphere. It seems to beget a peculiar sensitiveness to criticism. One magnificent head-mistress is "disgusted" that I do not know that in 1917 she was teaching exactly what I ask to have taught and taught properly in I937. Sonic day she may read my address in full, bring herself to study my innocent-looking diagram, and realise with a shock to how much foresight, modernity and religious novelty she had laid claim. Much capital is made of the fact that I hoped some day "1066 and all that" will be altogether forgotten in our schools. My assailants assume this to be an assertion that "1066 and all that" is what they are I teaching today. But is its They should read more carefully. The fact remains that the 1066 stuff is still going on in places now. And even when "1066 and all that" is left behind, it may still be a long long way to the necessary historical basis of a modern mind. A little stuff about "hunter peoples" and "plough peoples" may not settle the matter. I remain unconvinced of this alleged complete modernisation of history teaching. What alarms me most in this outburst of retorts is the tremendous self-satisfaction of so many of these acting educationists. I should hate to think it true that you can teach something to every man (or woman) except a schoolmaster (or schoolmistress). But you should see my mail and press cuttings this week. I remain absolutely sure that no proper treatment of the property-money conventions suitable for teaching has yet been achieved and I deny that any elementary education for the modern world state is possible until that is done. Nothing but twaddle and nonsense about property, money or economic control is being handed out to young people throughout the world. No picture of the economic world is given them. My magnificent schoolmistress, my hitting—out director of education, and all the rest of them, are in a state of self-protective hallucination about that.

I admit the extraordinary difficulty of creating a really modernised universal education, but I insist upon its Urgent necessity. In the course of the remaining discussions of the British Association it did become clear to us that we could not discuss education in vacuo. Education must have an objective and that objective must be the ideal of a community in which the educated person will live. Our Nazi and French visitors, Professor Levy and Mr. F. Horrabin, helped us to realise that. If the activities of the Educational Section of the British Association of this year did nothing else, they serve materially to show how inseparable is education from the general body of social science and theory.

Education and social existence are reciprocal. Its informative side has to be essentially social elucidation. So that the ideal teacher can never be a specialist; he has to have a working conception of the world as a whole into which his teaching fits. When I write or talk to teachers about the real magnitude of their task I am apt to feel like Max Beerbohm's caricature of Walt Whitman urging the American eagle to soar. It remained ruffled and inactive on its perch. Nevertheless for good or ill the future is in the hands of the teachers as it is in the hands of no other men and women, and the more this is recognised the more urgent our criticisms of them will have to be.

APPENDIX II. PALESTINE IN PROPORTION

PUBLISHED IN THE SUNDAY CHRONICLE,
OCTOBER 3RD, 1937)

THE other day I was talking to an assembly of teachers and scientific workers on the problem of getting the elements of a modern world outlook into the ordinary human mind during its all too brief years of schooling and initiation. I was not persuading nor exhorting; I was exposing my thoughts about one of the primary difficulties in the way of a World Pax which will save mankind from the destruction probable in putting the new wine of mechanical and biological power into the worn bottles of social and moral tradition. I dealt with the swiftness of life, the shortness of time available for learning and the lag and limitations of teaching.

In my survey of the minimum knowledge needed to make an efficient citizen of the world, I laid great stress upon history. It is the core of initiation. History explains the community to the individual, and when the community of interests and vital interaction has expanded to planetary dimensions, then nothing less than a clear and simplified world history is required as the framework of social ideas. The history of man becomes the common adventure of Everyman.

I deprecated the exaggerated importance attached to the

national history and to Bible history in western countries. I maintained that the Biblical account of the Creation and the Fall gave a false conception of man's place in this universe. I expressed the opinion that the historical foundation for world citizenship would be better laid if these partial histories were dealt with only in their proper relation to the general development of mankind. In particular I pointed out that Palestine and its people were a very insignificant part of the general picture. It was a side-show in the greater conflicts of Mesopotamia and Egypt. Nothing important, I said, ever began there or worked out there...

In saying that I felt that I was stating plain matter-of-fact to well-informed hearers. But it is not what I should have thought and said, forty years ago. And since the publication of my remarks, there have been a number of retorts and replies to my statement that have made me realise how widely and profoundly and by what imperceptible degrees, my estimate of this Jewish history has been changed since my early years and how many people still remain under my earlier persuasions. Long after I ceased to be a Christian, I was still obsessed by Palestine as a region of primary importance in the history of human development. I ranked it with Greece as a main source in human, moral and intellectual development. Most people still seem to do so. It may be interesting to state compactly why I have grown out Of that conviction.

Very largely it was through rereading the Bible after an i interlude of some years and with a fresh unprejudiced mind, that this change came about in my ideas. My maturer impression of that remarkable and various bale of literature which we call the Old Testament was that it had been patched a lot but very little falsified. Where falsification appeared, as in the number of hosts and slain in the Philistine bickerings, it was very naive, transparent and understandable falsification.

I was not impressed by the general magnificence of the prose, about which one still hears so much. There are some splendidly plain and vivid passages and interludes of great dignity and beauty, but the bulk of the English Bible sounds to me pedestrian translator's English, quite unworthy of the indiscriminate enthusiasm that has been poured out upon it. From their very diverse angles the books of the Bible have an entirely genuine flavour.

It is a collection; it is not a single book written ad hoc like the Koran. And the historical parts have the quality of honest history as well as the writers could tell it. Jewish history before the return from Babylon as the Bible gives it, is the unpretending story of a small barbaric people whose only gleam of prosperity was when Solomon served the purposes of Hiram by providing an alternative route to the Red Sea, and built his poor little temple out of the profits of porterage. Then indeed there comes a note of pride. It is very like the innocent pride of a Gold Coast negro whose chief has bought a motor-car. The prophetic books, it seems to me, reek of the political propaganda of the adjacent paymaster states and discuss issues dead two dozen centuries ago.

One has only to read the books of Ezra and Nehemiah to realise the real quality of the return of that miscellany of settlers from Babylon, a miscellany so dubious in its origins, so difficult to comb out. But a legend grew among these people of a Tremendous Past and of a Tremendous Promise. Solomon became a legend of wealth and wisdom, a proverb of superhuman splendour.

In the New Testament we hear of "Solomon in all his Glory." It was a glory like that of the Kings of Tara. When I remarked upon this essential littleness of Palestine I did not expect any modern churchmen to be shocked. But I brought upon myself the retort from the bishops of Exeter and Gloucester that I was obsessed by "mere size" and that I had no sense of spiritual values. My friend, Mr. Alfred Noyes, reminded me that many pumpkins were larger than men's heads, and what had I to say to that? But I had not talked merely of physical size. I had said that quite apart from size nothing of primary importance in human history was begun and nothing worked out in Palestine. That is I had already said quite definitely that Palestine was not a head but a pumpkin and a small one at that. A number of people protest. But, they say, surely the great network of modern Jewry began in Palestine and Christianity also began in Palestine! To which I answer, "I too thought that." We float in these ideas from our youth up. But have we not all taken the atmosphere of belief about us too uncritically? Are either of these ideas sound? I myself have travelled from a habit of unquestioning acquiescence to entire unbelief. May not others presently do the same? I do not believe that Palestine was the cradle of either Jewry or Christendom.

So far as the origin of the Jews is concerned, the greater probability seems to me that the Jewish idea was shaped mainly in Babylon and that the return to Judea was hardly more of a complete return and concentration than the Zionist return today. From its beginning the Jewish legend was a greater thing than Palestine, and from the first it was diffused among all the defeated communities of the Semitic-speaking world.

The synthesis of Jewry was not, I feel, very much anterior, if at all, to the Christian synthesis. It was a synthesis of Semitic-speaking peoples and not simply of Hebrews. It supplied a rallying idea to the Babylonian, the Carthaginian, the Phoenician, whose trading and financial methods were far in advance of those of the Medes, Persians, Greeks and Romans who had conquered them. It was a diffused trading community from the start.

Jewry was concentrated and given a special character far more by the Talmudic literature that gathered about the Old Testament collection, than by the Old Testament story itself Does anyone claim a Palestinian origin for the Talmud? I doubt if very much of the Bible itself was written in Palestine. I believe that in nine cases out of ten when the modern Jew goes back to wail about his unforgettable wrongs in Palestine he goes back to a country from which most of his ancestors never came.

When Paul started out on his earlier enterprise of purifying and consolidating Jewry before his change of front on the road to Damascus, he was on his way to a Semitic—a Jewish community there, and Semitic communities existed and Semitic controversies were discussed in nearly every centre of his extensive mission journeys. There was indeed a school of teachers in Jerusalem itself; but Gamaliel was of Babylonian origin and Hillel spent the better part of his life and learning in Babylon before he began to teach in Jerusalem. From the Bible itself and from the disappearance of Carthaginian, Phoenician and Babylonian national traditions simultaneously with the appearance of Jewish communities throughout the western world, communities innocent of Palestinian vines and fig trees and very experienced in commerce, I infer this synthetic origin of Jewry, Of course, if the reader is a believing Christian, then I suppose the crucifixion of Jesus of Nazareth at Jerusalem is the cardinal event of history. But evidently that crucifixion had to happen somewhere and just as Christian critics can charge me with being obsessed by mere size in my deprecation of Palestine, so I can charge them with being obsessed by mere locality. If the crucifixion has the importance attached to it by orthodox theologians then, unless my reading of theology is all wrong, it must be a universal and eternal and not a temporal and local event.

Moreover nowadays there is a considerable body of quite respectable atheists, theists and variously qualified Christians who do not find in that practically unquestionable historical event—I throw no doubt upon its actuality —the centre upon which all other events revolve. There has been a steady enlightenment upon the relations of Christian doctrine ceremony and practices to the preceding religions of Egypt, Western Asia and the Mediterranean, to the Egyptian trinity, to the Goddess Isis, to the blood redemption of Mithraism. In this great assembled fabric of symbols and ideas, the simple and subversive teachings of the man Jesus who was crucified for sedition in Jerusalem, play a not very essential part. Christianity, I imagine, or something very like it, would have come into existence, with all its disputes, divisions, heresies, protestantism and dissents, if there had been no Essenes, no Nazarenes and no crucified victim at all. It was a natural outcome of the stresses and confusions that rose from the impact of more barbaric and usually Aryan-speaking eonquerors, upon Egypt and upon the mainly Semitic-speaking civilisations, very much as Greek philosophy and art were the outcome of the parallel impact of the Hellenic peoples upon the Aegean cultural life. Old creeds lost their power and old usages their prestige. The temporarily suppressed civilisation sought new outlets. The urgency towards new forms of social and moral statement and adaptation was very great.

It was, I suppose, the advantage of the nexus of Semitic communities throughout the western world, that favoured the spread of Judaism and of the semi-Semitic Christianity that grew side by side with it rather than the diffusion of Persian religious inventions or Greek science and philosophy. It was an unpremeditated advantage. The thing happened so. And on that basis European mentality rests. We are all more or less saturated with this legendary distortion of historical fact. It makes us a little uncomfortable, we feel a slight shock when it is called in question.

Such is the conception of Jewish and Christian origins that has displaced the distortions of my early Low Church upbringing. It has robbed Palestine of every scrap of special significance for me and deprived those gigantic figures of my boyhood, Abraham, Isaac, Jacob and Moses of their cosmic importance altogether. They were local celebrities of a part of the world in which I have no particular interest. Once they towered to the sky. I want to get them and Palestine out of the way so that our children shall start with a better perspective of the world.

APPENDIX III. THE FALL IN AMERICA, 1937

PUBLISHED IN COLLIER'S, JANUARY 28TH, 1938

I SPENT October and November in America and saw Indian summer at its best, colours such as no other part of the world can boast, russet, glowing reds, keen bright yellows, soft green yellows, grey-blue, black-green, and skies of a serene magnificence. And the white American homes, grey-tiled, nestled brightly in that setting. In Philadelphia it rained, but it always rains in Philadelphia, and New York forgot itself for a day or so and blew and rained. Kansas City looked amazingly fine and handsome and I admired the parklands Henry Ford is laying out by Dearborn and the fine new pile of the Yale library, Gothic with a touch of skyscraper, and 2. great success at that. From the humming plane that took me from Detroit to Boston I looked out and saw Niagara foaming in the sunshine. I spent an instructive day at Flushing while the piledrivers hammered down hundred-foot trees into the mud, like a carpet-layer hammering tacks, preparing for the buildings that are to make New York's World Fair of 1939 the most wonderful ever. Everywhere colour, warmth, movement, vitality-and people talking about the new depression and possible war.

The depression that has struck America this autumn has been the most surprising thing in the world. It has been like the unaccountable failure of an engine. The wheels that had been spinning so busily slowed down until now the spokes are visible, and nobody on earth seems to know when they will pick up again or even whether they will pick up again. I came over to America once in the season of hope and hardship when President Franklin Roosevelt was newly in the White House. I thought then that he and Stalin were the most eventful persons in the world. They are both in their successes, such as they are, and in their human shortcomings, cardinal men. The old private-property money system was showing signs of age and an imminent breakdown. A new order was indicated as plainly in America as Russia. The New Deal, I assumed, was to be a real effective reconstruction of economic relationships and the Brain Trust was to get together and tell us how. I was particularly keen at that time to see and sample what I could of the Brain Trust, that improvised council of informed and constructive men which had to modernise and re-equip a staggering modern community. I found it a trifle incoherent. I went afterwards to Russia to talk to Gorky and Stalin about the absolute necessity for free discussion if a social order is to be effectively reconstituted. But Gorky I found grown old, fame-bitten and under the spell of Stalin, and Stalin, whom I liked, has never breathed free air in his life and did not know what it meant. When I revisited the President in 1935 things were asway and rather confusing. Now they are clearer. The New Deal was a magnificent promise, and it evoked a mighty volume of hope. Now that hope has been dissipated. Mass-hope is the most wonder-working gift that can come into the hands of a popular leader. The mass-hope of world peace at the end of the Great War, the mass-hope of the Russian revolution and the mass-hope of the New Deal were great winds of opportunity. But these great winds of opportunity do not wait for ships to be built or seamen to . learn navigation. They pass; they are not to be recalled. As I flew now over sunlit America and noted the traffics of life ebbing again below, and realised that that great capital of hope was nearly spent, I found the riddle of how people will behave, get past or stand up to, what is coming to them this year, a problem very difficult to contemplate in a sunlit manner. More and more of them will be short of food and shelter this winter— with no end in sight and nothing of the trustfulness that staved of disaster, perhaps only temporarily, in 1934.

Like hundreds of thousands of people I have had some sleepless nights over that riddle. It has been more and more vivid in my mind, since I wrote Anticipation: in 1900, that our world cannot struggle out of its present confusions and insufficiencies without a vigorous re-organisation of its knowledge, thought and will. Its universities, schools, books, newspapers, discussions and so on seem absurdly inadequate for the task of informing and holding together the mind of our modern world community. Something better has to be built up.

Nothing can be improvised now in time to save us from some extremely disagreeable experiences. The Flood is coming anyhow, and the alternative to despair is to build an ark. My other name is Noah, but I am like someone who plans an ark while the rain is actually beginning. This time I have been giving a lecture in a number of great cities about various possible educational expansions. I have been trying to interest influential people in schemes for knowledge organisation and I have been talking to teachers, professors, educationists—in considerable profusion.

I have never lectured before in America and only my real fanaticism about education made me attempt it at all. I liked it much more and it tired me much less, than I had anticipated, and my audiences abounded in pleasant young people who listened intelligently and asked intelligent questions. There were drawbacks—a processional hand-shaking, for example, and a disposition to lure the lecturing visitor by promises of tea and a quiet time, into large unsuspected assemblies where he is pressed to give an uncovenanted address. He is pushed through a door suddenly and there an ambushed audience is unmasked. It is not generally known in Europe—possibly I have been carried away by some misunderstanding—that in every considerable American city large gatherings of mature, prosperous, well-dressed women are in permanent session. They sit in wait, it seems, for any passing notoriety and having caught one insist on "a few words." This year they are all wearing black hats. These hats stick in my mind. Ultimately of the most varied shapes, the original theme seems to have been cylindrical, so that the general effect of an assembly of smart American womankind in 1937 is that of a dump of roughly treated black tin cans. The crazy irrelevance of this headgear on embattled middle-aged womanhood, is as essential a part of my memories of America this year, as the general disposition to discuss the depression and suggest nothing about it, and the still unstanched criticism of the President.

I talked to the President over a lunch tray and I told him how variously he was disapproved of and how incapable the opposition seemed to be of presenting a plausible alternative to him. It was our third meeting. I wrote of him some years ago as floating a little above the level of ordinary life. I find him floating more than ever. He seems to me to belong to the type of Lord Balfour, Lord Grey of Fallodon and Justice Holmes, great independent political figures, personally charming, Olympians detached from most of the urgencies of life, dealing in a large leisurely fashion with human stresses. The President is a skilled politician, as Holmes was a great lawyer, and Grey a bird watcher and fly-fisherman, but the quality their statesmanship has in common is its dignified amatcurishness. "Tell me," Balfour used to say, treating the other fellow as a professional whose business it was to know.

At the first convulsive intimations of failure in the economic machinery of America when Franklin Roosevelt came out as the saviour of his country, "Tell me" was in effect what he said and the Brain Trust was the confused response. I recalled his difficulties to him now, because I wanted to see how far he was a disappointed man and what sort of philosophy he had got out of it. The constitution had lain in wait for him, as every written constitution lies in wait for innovators. But his major insufficiency had been the quality of the aid and direction that the American universities and schools had given him. They hadn't told him, and instead of specialists they had yielded him oddities. The Brain Trust had proved very incalculable men. Men whom he had promoted had, he remarked, a trick of coming out against him. I pressed my obsession that America, like all the rest of the world, is in trouble because of its inadequate intellectual organisation. Men and women have been educated as competitive individuals and not as social collaborators and even at that the level of information has been low.

He agreed and began talking of certain experiments that had been made in the cultural development of Poughkeepsie county. It seemed to me an interesting and amiable exploitation of leisure, about as adequate to the urgencies of our contemporary situation, as polishing a brass button would be in a naval battle. I did not think him oblivious to the reality that America has to reconstruct its social life and cannot do so without a modernisation of education from top to bottom, but I got a very clear impression that he did not feel in the least responsible. He was not deeply interested in preparing for the future. That indifference is a common quality of the Olympian type.

The Olympian type assumes a competent civil service, but it cannot be troubled to make one. It takes the world as it finds it, and so the worst thing that can be charged against the President's administration is the continuation of the spoils system in the public services, for which I am told his close association with A. Farley is responsible. You cannot have safe administrators who do not feel safe.

We glanced at the possibility of a successor, but he did not seem to have any particular successor or type of successor in mind. We agreed that the danger of a world-wide war crisis would rise towards a maximum between 1939 and 1940 and he thought that by that time there should be some one younger, quicker, and better equipped to meet the urgencies of possible warfare without delay, in the White House. But he spoke of that rather as his own personal problem than Americas I left this autumnal president, feeling extremely autumnal myself and a day or so after I saw a play in New York, *I'd Rather Be Right*, in which I found a good-humoured confirmation of my own impression. It was a play about the American future, personified by a young couple who want to marry; the president, sympathetic but inadequate, was the principal character and the cabinet, the supreme court and so forth were presented under their own names. It bore marks of divergent suggestions, cuttings and rearrangements, but the genius and geniality of George M. Cohan made a delightful and sympathetic figure of the chief. The show is saturated with derisive affection. On my previous visits to America I had remarked that President Roosevelt was believed in enormously or hated intensely. The mood has changed. They like him now. They like him more than ever they did, and they believe in his magic no more.

What do they believe in? I varied my old stock question of his critics "What is your alternative?" to "You are in for a bad winter anyhow and what are you doing to prevent an indefinite prolongation of this decline?"

Lots of people just trust to Providence. "I have always come up before," they say as the drowning man said when he went down for the third time. So far as I could find out, that was the attitude of the old Republican guard. Big business in America appears to be completely bankrupt of political and social philosophy. Probably it never had any. It had simply a set of excuses for practices that were for a time extremely profitable and agreeable. It has over-capitalised the world, exhausted the land and stuck. Unhappily it sticks in the way. The only industrial leader who seems to be looking forward is the evergreen Henry Ford. I spent a congenial day with him at Dearborn and found him greatly concerned with growing the soya bean, which fixes free nitrogen and enriches the soil that every one else is exhausting. You can make everything from soup and biscuits to motor-car bodies and electric switches from the soya bean. But Ford, in addition to being a great inventive genius, is an individualist by habit and temperament and he stands outside the American scheme, a system in himself. He is like Science. He projects new things into the world, Ford cars which revolutionise the common roads and the common life of America, Ford tractors which set collectivisation afoot in Russia and now the limitless possibilities of soya. And like Science he has his political and social limitations. In the great American [conglomerate] he is like an island of something else. He does not like acquisitive finance and he does not like trade-unionism, but he does not know how to circumvent these two necessary outgrowths of our present competitive property-money system. He has his prejudice against Jewish particularism and his false estimates about the will for peace, but even in that prejudice and that false estimate there is maybe a gleam of prophetic foresight. A harder-thinking United States might have assimilated instead of isolating this outstanding imaginative genius. On the Left side of American affairs, strikes rather than ideas, increase and multiply. I was as much impressed by the number of pickets on the New York pavements as I was by the multitude of black can hats in the women's societies. I had heard a good deal about john L, Lewis as a coming man who Was going to do great things in politics. I met him and tried to find out how he thought the world was going. It reminded me of the distant past when I tried to get Clynes and Henderson and such-like lights of labour to tell me exactly what they thought they were going to do with the dear old British empire.

Maybe I misjudged him, but the impression I had was of a man, leonine according to the old senatorial model—he would look well in a cage with Senator Borah— and capable, but specialised largely in the purely labour issues of transport and mining. Anthracite and its rights and wrongs have entered into his soul. I aired my lifelong insistence that we and our world are all horribly at sixes and sevens mentally, and that first and foremost the world has to learn and think. He and my host denied hotly that the American Labour Party ignored education.

But I could not satisfy myself that what Labour means by education in America is anything more than an upward extension of the scholastic thing that is, qualified by a certain amount of training for political efficiency. I doubt if in America labour has got even so far as J.F. Horrabin of the British Plebs League or Laski and Strachey of the London Left Book Club, men who have evidently concluded long ago that equipment for an eternal class war is the sole end of human education. I do not believe that any benefit will accrue to America through the development of a special Labour Party in its political life. It is likely to be a heavy drag on intelligent reconstruction. As it has been in Britain. Labour parties have failed to become anything but trade-union parties and trade unionism is nothing more than the defensive organisation of the workers under a private capitalistic system. Its natural tactics are defensive and obstructive. It aims at shorter hours, better pay and a restraint upon dismissals. It is unable to imagine a new system. But a hundred years ago Karl Marx evolved a fantastic notion, partly from an inadequate analysis of British trade unionism and partly out of his inner consciousness, that the worker mass could become a mighty reconstructive force in the world. With no Blue Prints of what it was going to reconstruct. That would be the heresy of Utopianism. That delusion, embodied in communism and labour socialism, has undermined and checked the forces of science and creative liberalism for a century.

The British Labour leaders in power, showed themselves politicians within the containing politics of their time; they had neither the imagination nor the confidence in themselves to lay hands upon the universities, the diplomatic service, the foreign office and the monetary and financial organisations they found in being; they seem never to have heard of the gold standard until it hit them; and put to the test they did not even "nationalise" anything of importance. John L. Lewis may end in the White House as Ramsay Macdonald ended in royal favour, Clynes and Henderson in court costumes and Snowden and Webb in the house of lords. But "end" is the word for what any definitely Labour politician seems likely to do in the way of creative reconstruction.

I question indeed if the United States has sufficient time ahead to go through a phase of class politics at all. It has had all the possibilities of that worked out for it and ready for study—from Great Britain to Russia. It is an unprecedented country in its size, its freedom and its physical opportunities and I think the world has a right to expect something characteristic and original from it. Cannot it cut out that particular phase and get on?

The thing I found most hopeful among the falling equities and the falling leaves of this visit, was my occasional glimpses of the younger people. I saw more of them and I liked them better than I have ever done before. Every country nowadays' shows a contrast between the old and the young—a contrast in more than age; but here the contrast is astonishing. One still has the old boys with their stately, fatherly presences, uttering platitudes with an intensity of conviction unknown in the rest of the world, one is still introduced to a succession of presidential candidates with an irresistible suggestion about them of Tristram Shandy's bull—and THEN you meet the young.

In Henry James's *America Revisited* he tells of an encounter with a party of pre-war youth, drunken, noisy, coarsely sexual and hilariously irresponsible. I found very little of that, this journey. Instead I found a new generation, alert and interrogative. They have learnt about life in three courses of instruction. The disillusionment of the war made them pacifist. At first in rather a shallow fashion. They just proclaimed they were not to be humbugged into that sort of thing again. Dos Passos, that distinguished writer, has stuck at that stage, he is now a fossil from the first period. He proclaims that the Atlantic is too wide for air-raids, and has not yet discovered Mexico and South America nor the fact that America cannot keep her whole fleet in the Atlantic and the Pacific at the same time. But his juniors have taken these complications into account. Then before they could settle down into a qualified isolationism came the collapse that necessitated the New Deal. There again there was a tendency to think cheaply and there was a rush of uncritical communism, happily arrested—"happily" so far as America goes-by the Moscow trials and the Trotsky controversy, Now they seem to be facing the American problem in something like its real distinctness and complexity. They have to go further and reconstruct more fundamentally than Marx ever dreamt oh making new minds as well as a new world. I talked to a bunch at Harvard and I talked to a bunch at Yale and sampled individuals in the other places I visited. Cheap red paint is at a discount. I suppose that in a world of Tristram Shandy leaders, the phase of resentful insurgent communism was inevitable, but now in America you could put all the organised communists, rich undergraduates and genuine proletarians together, into a third-rate town and still have houses to let.

Reconstruction through socialisation, strenuous educational work to build up a competent receiver for bankrupt and expropriated public utilities, a steady development of a loyal civil

service, freed from—"Farleyism" seems to be the word I want—
and after the harsh winter that must surely follow this present
Fall, a new spring may break upon the world from America.
Through its renascent young people.

Two young men—for so I speak of men in the early forties,
nowadays—produced a vivid the upon me, the presidents of
Harvard and Yale. They are something new in my experience of
Americans, something fresh, clear, frank and simple. President
Conant of Harvard, for example, is a very distinguished chemist
indeed; he has the balanced lucid mind of the research addict, and
he is deliberately fuming from physical science to educational
and administrative work. Every one speaks well of him. Ever
since I met him I have been asking whether there are more like
him and why he is not in the running for the White House. "He
has been talked of," I was told, "but—"

"Well?"

"You see we've already had one college president there—
Wilson."

It seems an inadequate excuse. For Wilson was a professor of
history, that is to say, a man trained to be unscientific.

Indian summer still reigned as the *Queen Mary* with a sort
of lazy swiftness pulled out from dock. I watched the great grey
and brown and amber masses, cliffs and pinnacles of middle and
then of lower New York, soften in the twilight and light up. What
a spectacle it is!, Such towering achievement and so little finality.
How much more vitally unfinished than the contour of dear old
Saint Paul's, brooding like a shapely episcopalian hen over the
futile uneasiness of London! America seems to have a limitless
capacity for scrapping and beginning again.

APPENDIX IV. TRANSATLANTIC
MISUNDERSTANDINGS

PUBLISHED IN LIBERTY, JANUARY 15TH, 1938

I HAVE recently been going to and fro in America, talking to all sorts of people and hearing all kinds of opinions. I am one of those realists who refuse to be blinkered by current political institutions. I belong to that great and growing community which has a common literature and a common language in English. There has been a revolution in communications during the past hundred years, a virtual abolition of distance, which makes our old separations preposterous. The English-speaking states and nations scattered about the world are no longer divergent. They are coming together again, in their thoughts, knowledge, interests and purposes. They have a common future.

I believe profoundly in the synthetic power of a common idiom of thought and expression. Given a free movement of books, papers, radio beams and the like, this is a unifying force that will triumph ultimately over every form of particularism, nationalism and imaginative antagonism. It is bringing the Californian into clearer and closer understanding with the Kentishman, the New Yorker, the Scot, the Maori, the Afrikander, the Canadian, the Welshman, the Anglo-Indian and the Eurasian. I doubt if the green barriers of a censorship and an artificial language will suffice to keep the Irish Free State out of our world-wide coalescence for long. Irishmen are not going to live content talking faked Erse in a small damp island when they have for wit and abuse and the whole English-speaking world at their disposal. I refuse to call myself foreign or alien among any people who speak, read and think in English. It is because they are less reasonable today than ever they were, that I find the resistances to this greater fusion, increasingly remarkable. There are such resistances, one must admit. There are still all sorts of queer and out-of-date frictions and obstructions in the confluence of this new world-wide English-speaking community of ours. This is particularly evident between the United States and Britain. Partly this is due to bad tricks of manner and bad habits of mind on the part of the British which irritate many types of Americans excessively; partly it is due to deep-rooted traditions of distrust and antagonism which have long since ceased to have any practical justification.

Almost all English people forget that so far as origins go the population of the United States never was wholly British. They disregard the early Dutch, Swedish, French and other non-British settlers who slowly adopted the English language and traditions before the War of Independence and the immense infusion of the later immigrations, immigrants who had no earthly reason for Anglomania. These people accepted English and the forms of English thought, they amalgamated with the American school of the English-speaking culture, many of them transferred their ancestors to the *Mayflower* and acquired a kind of pre-natal pride in the Battle of Bunker Hill, yet steadily and particularly since the beginning of this century, their distinctive qualities have manifested themselves in a greater variety, a wider range, in American imaginative literature and a broader, more enterprising quality in American thought. For long Boston remained more English than England, but forty years ago my generation hailed the intellectual revolt against Boston (and refined Anglicanism) first in a sprinkling of isolated writers from Whitman to Stephen Crane, and then in a comprehensive expansion and release, which has now made American intellectual life not so much a continuation as a vast extension and Europeanisation of English culture. The annexations have been enormous. The English element has played the part, not so much of a direction as a flux. These are matters all too frequently ignored by English people. The thing has happened in history before. just as American education has oriented itself to the War of Independence, in which not one in a score of living Americans had an active ancestor, so the great Phoenician, Babylonian and Carthaginian populations acquired and entered into that tradition of Judea which holds together the comity of Judaism today. Probably, too, under the empire, millions of Roman citizens from York to Egypt without a drop of Roman blood in their veins felt a personal pride in Romulus and Remus. Assimilation is far easier and more powerful than inheritance. It is the linguistic link and the progressive development of common ideas and understandings that are destined to hold the English-speaking community together in the future. Britain never was the mother country of the United States, and the first thing an English visitor to America should do is to get rid of that illusion. Long years ago, a century ago, one of the great New England writers of that time—was it Lowell?—wrote a stinging paper on "a certain air of patronage in foreigners." It was an excellent reproof to visiting Europeans and particularly to Englishmen who came over to see America as a new young country, betrayed a sort of upper-form attitude to the new boys, and were quite unable to realise that a man of forty in America is just as old as a man of forty in England, has seen just as much of life and counts as many ancestors between himself and Adam. The political association of Britain and America lasted hardly a couple of centuries. But it set a perennial stamp upon their destinies. From the beginning of their separation there has been a peculiar mutual irritation between Britain and America. It has in it something of the vexation of a divorced couple who still resentfully care for things they had in common. "Think what the Empire would have been if you had stuck to us," is the unspoken accusation of the ordinary Britisher. "Why did you make things so that we had to break?" says the American. "It was your fault," say both of them, "and my side was in the right." Maybe you will think that is now ancient history. It is not. It is as alive today as it was a century ago. The mighty growth of America makes the British none the less regretful for the loss of a common world outlook. They paid too much for clear old George the Third. It was the Hanoverian monarchy, the hungry exploitation of America by the British governing class and the narrow traditions of the Foreign Office that caused that breach. Plebeian Britain was on the side of the colonies in the War of Independence— John Wilkes was very brother to Tom Paine—and from Gibbon's *Decline and Fall* (1776-88) to Winwood Reade's *Martyrdom of Man* (1872) you can find the liveliest anticipations in English literature of the role of North America as the hope and refuge of Western culture and freedom. Nothing is forgotten so rapidly as very recent history, and few people nowadays realise how near Britain came a century ago to following the United States along the path to republicanism and to emancipation from an aristocratic ruling class. Joseph Chamberlain, the father of the present Prime Minister, was a republican outright, but nowadays we forget his talk of ransom—his New Deal phase. Both countries are amazingly ignorant of each other's history during the early nineteenth-century period—they were too busy in other directions to observe each other closely—and few Americans seem to realise the struggle that went on in England during their civil war between the ruling class, which looked forward with pleasant resentment to the break up of the Union, and the liberal British, who instinctively realised that the strength and freedom of America were bound up with their own. To this day relics of that old upper-class resentment lingers. The Bourbons of the court and the diplomatic services still betray a lurking patronage, depreciation and a weak vindictiveness towards America. "Society" is in politics in Britain and out of it in America, but the economic, social and political trends in both systems are alike, their massive interests and mental dispositions converge, and however these retrospective elements may falsify our superficial relationships the enduring disposition of the generality in Britain towards America and the drive behind the national policy are, and always have been, co-operative.

But balancing the retrospective resentment of the British ruling class, there is on the other side a whole system of estrangement between these great populations arising primarily out of the too-cherished tradition of the War of Independence. For several generations that great revolt formed the basis of historical instruction in America. The bridge at Concord, the Boston Tea Party were exaggerated gigantically. The new immigrants took them over. The incoming German or Eastern European learnt to boil with indignation at the employment of Hessian troopers by the Hanoverian king. The inevitable misdemeanours on the part of the British were charged up against them in the true spirit of war propaganda. In the war of 1812 they burnt the Capitol. This was not all. All that might have been forgotten. But the newly separated states developed a financial organisation only very slowly. They were not industrialised for nearly a hundred years and they had to be industrialised by foreign capital—which in those days meant British capital. And they developed a diplomatic service belatedly. Because of this an inferiority complex towards the British arose, quite naturally, but one that has long since lost any material justification. Wall Street as it grew seemed to the Americans generally very like a British foothold in their economy. America felt that it was entangled monetarily, sold out to London, mysteriously out-manoeuvred in its foreign relations.

America grew, a unified continental mass. Her population is now two-thirds of the English-speaking world, while the British find themselves responsible for an empire which spreads about the planet, the most vulnerable, the most entangled, the most threatened of all existing political systems. Its foreign policy has become now an almost pathetic self-protective opportunism. But to the American this empire is still a greedy octopus, to be distrusted systematically. It has kept an unguarded frontier to the north of him since the separation, nevertheless it is to be distrusted. Its fleets have never threatened America and for fifty years at least Americans have had no uneasiness on their eastern coasts very largely on account of Great Britain, nevertheless it is to be distrusted. Wherever the Union Jack goes it takes the English-speaking American commercial traveller and a friendly banking system with it, nevertheless it is to be distrusted. Some mysterious undermining process is believed to be going on. The phrase adopted, since the War to express this profound distrust has been "British Propaganda."

Suspicion of British propaganda has become a mania. Old ladies in the middle west look under the bed at nights for British propaganda. At home, I am a persona most distinctly non gram with the court, the church, the public schools, the universities, the whole Anglican system; but when I go to America I am discovered to be a British propagandist. America in her serene path to abundance and happiness, is being lured into another war—for the sake of the British. That is the story. America has no interest in the welfare of China, it seems, because the British have great investments there. But we tempt her in. We invented the Pacific coast on her. It is British propaganda to suggest that America has some slight interest in the destinies of the Spanish and the Portuguese-speaking worlds. To say that isolation from the rest of the world's affairs is a rather cowardly and altogether impracticable ideal for America nowadays, is British propaganda.

There is a copious literature of propaganda against this imaginary British propaganda. The other day I read a book by Mr. Quincy Howe, England Experts Every American to do his Duty, which is quite typical of the methods of the campaign of distrust and estrangement. The history of Anglo-American relations is combed through to present America as the simpleminded cat's-paw of British cunning. This sort of thing crops up everywhere. I found it in the questions put to me after my lectures; I met it in conversation. "You British" they began. Though I had come to America to talk about educational reorganisation and not about international affairs. A hostile power preparing for a conflict with England solo or America solo could not cultivate a crippling breach of their natural alliance more efficiently than this probably quite spontaneous misinterpretation of the British drift. It may prove a very disastrous thing for the entire English-speaking world.

APPENDIX V. THE ENGLISH-SPEAKING WORLD: "AS I SEE IT"

BROADCAST TALK DELIVERED DECEMBER 1ST, 1937

I FIND myself on the air for the Empire broadcasting service—free to speak for a quarter of an hour on practically any subject that occurs to me—under this most liberating title of As I See it. I suggest that, As I Think about it, would have been a better title. What I see is a brightly lit desk, a lamp, a microphone in a pleasantly furnished room—and a listener, for I never talk for broadcasting without a real live listener actually in the room with me—but what I am thinking about is a great number of listeners, some alone, some in groups, in all sorts of rooms and places, all round the world. We are, I guess, an extremely various and scattered lot indeed, race, religion, colour, age. We have probably only one thing in common. Which is that we speak, write and understand English.

I want to talk about ourselves and the community to which we belong. I see that as a tremendous world brotherhood full of possibilities and full of promise for the hope, the peace, the common understanding of all mankind.

I have been asked by the Empire Broadcasting Service to make this talk, but it is, you must understand, a quite uncontrolled talk or I would not give it. I hold no brief for the Empire as such;

it is a complex of political arrangements, which are constantly changing and will continue to change. Widely as it extends, it does not include the larger part of this English-speaking brotherhood of ours, which is to my mind something infinitely more real, more important and more permanent. I am talking reality—not propaganda.

I spent this autumn in the United States. I was lecturing there about intellectual organisation in schools and universities and I talked with all sorts of people from the President and Mr. Henry Ford downward. We all talked the same language, in the same idiom of thought. We understood each other pretty thoroughly. Yet we are drawn from the most diverse sources. It is a common mistake among English people to suppose that Americans are just English people transplanted. But from the very beginning the United States were of diverse origin. The Swedes, the Dutch, Germans, the French in Louisiana, the Spanish i.n California, were there as soon as the New Englanders, long before the War of Independence. Afterwards there was an enormous influx of Eastern Europeans. And again in the British Empire itself, there is a great assembly of once alien peoples drawn together into a common interchange—from the Eskimo of the Labrador Coast to the Maori of New Zealand. But the English language has amalgamated—or is amalgamating-all these elements into a great cosmopolis, whose citizens can write to each other, read and understand each other, speak freely and plainly to each other, exchange, acquire and modify ideas with a minimum of difficulty. Once or twice before in history there has been such a synthesis in the Latin-speaking world, in the Semitic-speaking world, but never on such a vast scale as in this English-speaking world in which we live and think today. And English has never been forced upon these multitudes who speak it now, they were never subdued to it or humiliated by it, they have taken it up freely and they use it of their own good-will because it serves them best. Now a thing that impresses me greatly, it seems to me one of the most important things in our present world, is that this English-speaking community is not breaking up and does not look like breaking up, into different languages. In the past that sort of thing did occur. Latin, as you know, broke up into French, Italian, Castilian, Catalan and a multitude of minor dialects. But since then a vast change has occurred in the conditions of human life; the forces of separation have been dwindling, the forces that bring us nearer to one another have been increasing enormously, the printed word, books, newspapers, the talking movie, the radio, increasing travel, increasing trade, now forbid dispersal. History has gone into reverse. Instead of being scattered about the earth and forgetting one another, a thing which happened to the Aryan speakers and the Mongolian speakers of the past, we English speakers are being drawn together and learning more and more about each other. This reversal of the old order of things has been going on ever since the steamship and the railways appeared, a century ago. It goes on faster and faster. In the past new dialects were continually appearing; new dialects are disappearing. The curse of Babel has been lifted from over three hundred million people. This coming together is a new thing in human experience. And having got this unprecedented instrument of thought spread all about the world, a net of understanding, what are we English speakers doing with it to get the best out of it? Are we getting the best out of it?

Are we growing into one mighty community of ideas and sympathies and help and peace as rapidly as we might do? I do not think we are. Something, I admit, is being done to realise the tremendous opportunity of the world-wide spreading of the English language, but nothing like what might be done, if we grasped our possibilities to the full.

Let me tell you as briefly as I can one or two of the things that might be done to make this great gift of a common language better worth while. They are things every one of us in this talk tonight can set about demanding at once. You can write to your representative or member of parliament about them before you go to bed.

First about books. Nothing can pull our minds together as powerfully as books. We all want to read books according to our interests and habits. We find them so dear to buy or so difficult to borrow that most of us cannot read half the books we hear about. And three-quarters of what books there are, we never hear about at all. This is true even here in London. Here I am on the telephone to well-stored book shops and all sorts of people from whom I can get advice. Even so it is true here. But a majority of my listeners tonight may be living in parts of the Empire far away from the centres of book distribution. Mentally many of them must suffer the torments of Tantalus. They perceive there is a great and refreshing flood of ideas, imaginative, informative, matter, fantasy, poetical invention flowing through the English world and they Call get only just a splash or so of it to their thirsty lips. In Great Britain in the larger towns you can buy a fair selection of the best books published, even quite new books, for from sixpence to a shilling. But in America there are no really cheap books and the great mass of the workers and poor people there, never read books at all. There are public libraries, of course, where you can wait for books for quite a long time. Most of our 300,000,000 English speakers, through no fault of their own, read nothing better than a few odd books that chance to come their way. They never acquire the habit of systematic book reading. English, which should be the key of all human thought and knowledge, is for them the key to a non-existent door.

The reading, thinking section, the book-reading section, of the Empire probably does not number a million all told. The rest either read newspapers or do not read at all.

Now before you blame the public or the schools or the booksellers for this immense illiteracy, this great mental underdevelopment, consider the difficulties of sending books about. Try sending a book, a good fat book, half-way round the world and see what it costs you. You will realise that a special low postal rate for books and parcels of books, a special preference rate, a rate to encourage the sending of books, is one of the first things necessary before we can begin to realise the full cultural promise of our widespread English tongue. It is a matter that should concern every Ministry of Education. Does it?

Given such rates you'll soon rind every publisher in the world building bigger printing plants and selling books for sixpence—almost as soon as they are issued. But book postage is not considered a public service. It is made a source of revenue and until people like ourselves who read and listen in and want to know begin to make a fuss about it, matters will remain very much as they are.

Cheap good books—and next comes the problem of how to hear of them—so that we may—from the ends of the earth—order the ones we really want and spend our sixpences properly. Well, probably half my hearers have never heard of what is called documentation, and they think bibliography is something remote and scholastic and all that sort of thing. But really it is nothing more or less than indexing all that has been written in the world, so that you can find out quickly and surely what has been done, by whom, and under what title. Don't you want to know that? And do you know it? There are hundreds of clever people working out methods of indexing and in a little while it will be quite possible to print and keep up-to-date bibliographies, lists of all the best books, in every great group of subjects in the world. It would be as easy to keep up such bibliography as it is to keep up the issue of railway time-tables. The cost of producing these book guides need not be very much greater than the cost of producing those time-tables. I doubt if today a hundred thousand of us use any bibliographies at all. What is the good of reading unless you know what books to read? Bibliographies ought to lie about in every educated household.

And another thing which we English speakers have a right to ask for, considering what a vast multitude we are and all that we might be, and that is a general summary of contemporary knowledge and ideas, a real modern, adequate Encyclopaedia, kept up to date and available for the use of any one. That would hold us all together as nothing else would do. We should all be of a mind and nothing on earth would have the strength to stand against our thinking. But is there anything of the sort? No. The latest Encyclopaedia in my study is dated 1929—eight years old—and it is a very imperfect performance at that. Very old-fashioned. Very little better than the Encyclopaedias of a hundred years ago. Discovery and invention have been going on vigorously for the past eight years—but how am I to learn quickly about that new stuff? There is not a sign of a new one in sight. Does any one care—any of our education departments? Not a rap. The French just now—in spite of threats of war, in spite of great financial difficulties are making a new and a very admirably planned Encyclopaedia. You may think an Encyclopaedia is something only rich people can afford to buy. It ought not to be. If you can afford a radio set, if you can afford a motor-car, surely you can afford a summary of human thought and knowledge. Encyclopaedias need not be as dear as they are, any more than books or bibliographies. Cheaper books, handy bibliographies, a great encyclopaedia, our English-speaking world needs all these things. When automobiles first came along, they seemed likely to become a rich man's monopoly. They cost upwards of £1,000. Henry Ford altered all that. He put the poor man on the road. We want a Henry Ford today to modernise the distribution of knowledge, make good knowledge cheap and easy in this still very ignorant, ill-educated, ill-served English-speaking world of ours. Which might be the greatest power on earth for the consolidation of humanity and the establishing of an enduring creative Pax for all mankind.

My quarter of an hour is at an end. I haven't said half of what I would like to say. But if I have made you a little discontented with what we are doing with this precious inheritance of ours— English, I shall not have used this bit of time in vain.

THE END

Printed in Great Britain
by Amazon